Edgar Thurston, Madras, India Central Museum

Notes on the Pearl and Chank Fisheries

And marine fauna of the Gulf of Manaar

Edgar Thurston, Madras, India Central Museum

Notes on the Pearl and Chank Fisheries
And marine fauna of the Gulf of Manaar

ISBN/EAN: 9783337272845

Printed in Europe, USA, Canada, Australia, Japan

Cover: Foto ©berggeist007 / pixelio.de

More available books at **www.hansebooks.com**

GOVERNMENT CENTRAL MUSEUM, MADRAS.

NOTES

ON THE

PEARL AND CHANK FISHERIES

AND

MARINE FAUNA OF THE GULF OF MANAAR.

BY

EDGAR THURSTON, C.M.Z.S., &C.,
SUPERINTENDENT, GOVERNMENT CENTRAL MUSEUM.

MADRAS:
PRINTED BY THE SUPERINTENDENT, GOVT. PRESS.
1890.

I.—TUTICORIN PEARL FISHERY.

II.—PEARLS OF MYTILUS AND PLACUNA.

III.—TUTICORIN CHANK FISHERY.

IV.—CEYLON PEARL FISHERY.

V.—RÁMÉSVARAM ISLAND.

VI.—MARINE FAUNA OF THE GULF OF MANAAR.

VII.—INSPECTION OF CEYLON PEARL BANKS.

> "Know you, perchance, how that poor formless wretch—
> The Oyster—gems his shallow moonlit chalice ?
> Where the shell irks him, or the sea-sand frets,
> This lovely lustre on his grief."
>
> *Edwin Arnold.*

I.--THE TUTICORIN PEARL FISHERY.

TUTICORIN, the "scattered town," situated on the south-west coast of the Gulf of Mannar, from which the Madras Government pearl fishery is conducted, is, according to Sir Edwin Arnold,[1] " a sandy maritime little place, which fishes a few pearls, produces and sells the great pink conch shells, exports rice and baskets, and is surrounded on the landside by a wilderness of cocoa and palmyra palms."

Summed up in these few words, Tuticorin does not appear the important place which, in spite of its lowly appearance when viewed from the sea and the apparent torpor which reveals itself to the casual visitor, it is in reality, not only as a medium of communication between Tinnevelly and Ceylon, to and from which hosts of coolies are transported in the course of every year, but as being an important mercantile centre for the shipment of Tinnevelly cotton, jaggery, onions, chillies, &c.

With respect to the shipment of jaggery, I was told, during a recent visit to Tuticorin, that, during the seasons at which jelly-fish abound in the muddy surface water of the Tuticorin harbour, so great is the dread of their sting, that coolies, engaged in carrying loads of palmyra jaggery on their heads through the shallow water to the cargo boats, have been known to refuse to enter the water until a track, free from jelly-fish, was cleared for them by two canoes dragging a net between them.

Tuticorin is, indeed, " an abominable place to land at," and it is unfortunate that it is ordained by nature that large vessels shall not approach nearer to the shore than a distance of six miles or thereabouts, being compelled, with due regard for their safety, to lie at anchor outside Hare Island, one of

---

[1] *India Re-visited*, 1887.

a number of coral-girt islands in the neighbourhood, where hares and partridges may be shot, and sluggish *Holuthurians* captured in abundance at low tide as they lie impassive on the sandy shore, which is strewed with broken coral fragments, detached by wave-action from the neighbouring reef, and riddled with the burrows of nimble Ocypods (*O. macrocera* and *O. ceratophthalma*.) The habits of the latter species of crustacean are well described by Sir J. Emerson Tennent, who writes [1] :—

" The *ocypode* burrows in the dry soil, making deep excavations, bringing up literally armfuls of sand, which, with a spring in the air, and employing its other limbs, it jerks far from its burrows, distributing it in a circle to the distance of several feet. So inconvenient are the operations of these industrious pests that men are kept constantly employed at Colombo in filling up the holes formed by them on the surface of the Galle face. This, the only equestrian promenade of the capital, is so infested by these active little creatures that accidents often occur through horses stumbling in their troublesome excavations."

Not far from the north end of the town of Tuticorin, on the sandy shore, are the kilns, in which corals, coarse mollusc shells (*Ostræa, Venus, Cardium*, &c.), and melobesian nodules (calcareous algæ) are burned and converted into *chunám*,[2] *i.e.*, prepared lime used for building purposes, and by natives for chewing with betel. A native informs me that in the Bombay and Bengal Presidencies and in the North-Western Provinces pearls are bought by wealthy natives to be used instead of chunám with the betel. In India relations and friends put some rice into the mouth of the dead before cremation, but in China seed pearls are used for the same purpose.

During my visit to Tuticorin in 1887, I used to watch, almost daily, grand, massive blocks of *Porites, Astræa* and various species of other stony coral genera, being brought in canoes from the reefs and thrown into the ground to form the foundation of the new cotton mills, which, in consequence, bear the name of the Coral Mills.

Lecturing at the Royal Institution [3] on the "Structure, Origin, and Distribution of Coral Reefs and Islands,"

---

[1] *Sketches of the Natural History of Ceylon*, 1861.
[2] The familiar house frog (*Rhacophorus maculatus*) of Madras is popularly known as the "chunám frog" from its habit of sticking on to the chunám walls of dwelling houses.
[3] Friday, March 16, 1888.

Mr. John Murray stated that "if we except Bermuda and one or two other outlying reefs where the temperature may occasionally fall to 66° Fahr. or 64° Fahr., it may be said that reefs are never found where the surface temperature of the water, at any time of the year, sinks below 70° Fahr., and where the annual range is greater than 12° Fahr. In typical coral reef regions, however, the temperature is higher and the range much less." No regular series of records of the temperature of the water in the coral-bearing Gulf of Manaar has as yet been made. The surface temperature, which I recorded from time to time during my visit to Rámésvaram island in the latter half of July 1888, varied from 79° Fahr. to 91° Fahr. between the hours of 7 A.M. and 6 P.M.

The following table shows the temperature range of Tuticorin during the year 1887, the readings being taken in the shade at 10 A.M. and 4 P.M. :—

	Range.	Min.	Max.
January	9°	75°	84°
February	6°	78°	84°
March	9°	80°	89°
April	12°	79°	91°
May	13°	83°	96°
June	9°	86°	95°
July	10°	86°	96°
August	11°	84°	95°
September	9°	85°	94°
October	6°	80°	86°
November	7°	79°	86°
December	11°	75°	86°

Tuticorin has been celebrated for its pearl fishery from a remote date, and, as regards comparatively modern times, Friar Jordanus, a missionary bishop, who visited India about the year 1330, tells us that as many as 8,000 boats were then engaged in the pearl fisheries of Tinnevelly and Ceylon.[1]

In more recent times the fishery has been conducted, successively, by the Portuguese, the Dutch, and the English. The following excellent description by Martin of the pearl fishery in the year 1700, during the Dutch occupation of Tuticorin, shows that the method of fishing adopted at that time agrees, in its essential characters, with that which is in vogue at the present day :—

"In the early part of the year the Dutch sent out ten or twelve vessels in different directions to test the localities in which

---
[1] Streeter, *Pearls and Pearling Life*, 1886.

it appeared desirable that the fishery of the year should be carried on; and from each vessel a few divers were let down who brought up each a few thousand oysters, which were heaped upon the shore in separate heaps of a thousand each, opened, and examined. If the pearls found in each heap were found by the appraisers to be worth an *écu* or more, the beds from which the oysters were taken were held to be capable of yielding a rich harvest; if they were worth no more than thirty sous, the beds were considered unlikely to yield a profit over and above the expense of working them. As soon as the testing was completed it was publicly announced either that there would or that there would not be a fishery that year. In the former case enormous crowds of people assembled on the coast on the day appointed for the commencement of the fishery; traders came there with wares of all kinds; the roadstead was crowded with shipping; drums were beaten, and muskets fired; and everywhere the greatest excitement prevailed until the Dutch Commissioners arrived from Colombo with great pomp and ordered the proceedings to be opened with a salute of cannon. Immediately afterwards the fishing vessels all weighed anchor and stood out to sea, preceded by two large Dutch sloops, which in due time drew off to the right and left and marked the limits of the fishery, and when each vessel reached its place, half of its complement of divers plunged into the sea, each with a heavy stone tied to his feet to make him sink rapidly, and furnished with a sack into which to put his oysters, and having a rope tied round his body, the end of which was passed round a pulley and held by some of the boatmen. Thus equipped, the diver plunged in, and on reaching the bottom, filled his sack with oysters until his breath failed, when he pulled a string with which he was provided, and, the signal being perceived by the boatmen above, he was forthwith hauled up by the rope, together with his sack of oysters. No artificial appliances of any kind were used to enable the men to stay under water for long periods; they were accustomed to the work almost from infancy, and consequently did it easily and well. Some were more skilful and lasting than others, and it was usual to pay them in proportion to their powers, a practice which led to much emulation and occasionally to fatal results. Anxious to outdo all his fellows, a diver would sometimes persist in collecting until he was too weak to pull the string, and would be drawn up at last half or quite drowned, and very often a greedy man would attack and rob a successful neighbour under water; and instances were known in which divers who had been thus treated took down knives, and murdered their plunderers at the bottom of the sea. As soon as all the first set of divers had come up, and their takings had been examined and thrown into the hold, the second set went down. After an interval, the first set dived again, and after them the second; and so on turn by turn. The work was very exhausting, and the strongest man could not dive oftener than seven or eight times in a day, so that the day's diving was finished always before noon.

"The diving over, the vessels returned to the coast and discharged their cargoes; and the oysters were all thrown into a kind of park, and left for two or three days, at the end of which they opened and disclosed their treasures. The pearls, having been extracted from the shells and carefully washed, were placed in a metal receptacle containing some five or six colanders of graduated sizes, which were fitted one into another so as to leave a space between the bottoms of every two, and were pierced with holes of varying sizes, that which had the largest holes being the topmost colander, and that which had the smallest being the undermost. When dropped into colander No. 1, all but the very finest pearls fell through into No. 2, and most of them passed into Nos. 3, 4, and 5; whilst the smallest of all, the seeds, were strained off into the receptacle at the bottom. When all had staid in their proper colanders, they were classified and valued accordingly. The largest or those of the first class were the most valuable, and it is expressly stated in the letter from which this information is extracted that the value of any given pearl was appraised almost exclusively with reference to its size, and was held to be affected but little by its shape and lustre. The valuation over, the Dutch generally bought the finest pearls. They considered that they had a right of pre-emption. At the same time they did not compel individuals to sell if unwilling. All the pearls taken on the first day belonged by express reservation to the King or to the Sétupati according as the place of their taking lay off the coasts of the one or the other. The Dutch did not, as was often asserted, claim the pearls taken on the second day. They had other and more certain modes of making profit, of which the very best was to bring plenty of cash into a market where cash was not very plentiful, and so enable themselves to purchase at very easy prices. The amount of oysters found in different years varied infinitely. Some years the divers had only to pick up as fast as they were able and as long as they could keep under water; in others they could only find a few here and there. In 1700 the testing was most encouraging, and an unusually large number of boat-owners took out licenses to fish; but the season proved most disastrous. Only a few thousands were taken on the first day by all the divers together, and a day or two afterwards not a single oyster could be found. It was supposed by many that strong undercurrents had suddenly set in owing to some unknown cause. Whatever the cause, the results of the failure were most ruinous. Several merchants had advanced large sums of money to the boat-owners on speculation, which were, of course, lost. The boat-owners had in like manner advanced money to the divers and others, and they also lost their money."

In the present century the following fisheries have taken place :—

1822 .. .. profit	£13,000	
1830 .. .. do.	£10.000	
1860-62 .. .. .. do.	Rs.	379,297
1889 .. .. do.	,,	158,483

As to the cause of the failure of the pearl oysters to reach maturity on the banks in large numbers, in recent times, except after long intervals, I, for my part, confess my ignorance. Whether the baneful influence of the mollusca known locally as súran (*Modiola*, sp.) and killikay (*Avicula*, sp.), the ravages of rays (*Trygon*, &c.) and file-fishes (*Balistes*), poaching, the deepening of the Pámban Channel, or currents are responsible for the non-production of an abundant crop of adult pearl-producing oysters during more than a quarter of a century, it would be impossible to decide, until our knowledge of the conditions under which the pearl oysters live is much more precise than it is at present.

The argument that the failure of the pearl fishery is due to poaching is, from time to time, brought forward; but, as Mr. H. S. Thomas wisely and characteristically remarks [1] " the whole system of the fishery has been carefully arranged, so that everyone in any way connected with it has a personal stake in preventing poaching, and oyster poaching is not a thing that can be done in the night; it must be carried out in broad daylight; and, to be worth doing at all, it must be done on a large scale. Ten thousand oysters cannot be put in one's pocket like a rabbit, nor are there express trains and game-shops to take them. Every single oyster has to be manipulated, and it is only the few best that can be felt at once with the finger, and the usual way is to allow the oyster to rot and wash away from the pearl. Oysters could not be consigned fresh in boxes or hampers by rail to distant confederates; they could not even be landed without its becoming known ; and, if known, every one is interested in informing the Government officer and stopping poaching." I cannot, however, refrain from quoting the following touching description of an ideal poach in a recent pamphlet :—

" Mutukuruppan and Kallymuttu are two fishermen brothers : they start out after their cold rice, ostentatiously to

---

[1] Vide *Report on Pearl Fisheries and Chank Fisheries*, 1884, by the Hon. Mr. H. S. Thomas.

get their lines ready in their canoe, and paddle away to their fishing ground; there they drop their stone anchor: presently one observes that it is warm and he would like a bathe; over the side he goes down by his mooring rope to see what the bottom is like. He brings up a handful of oysters and gives them to Thamby; then Thamby thinks he would like a bathe, and he goes down also, and brings up a fist full. When they are tired they get back into the canoe and open their spoils, taking out what pearls they can find, and pitching the shells back into the sea. This sort of thing goes on day after day and year after year up and down the coast, and this will partially account for the dead shells so often found on the banks. Is it to be wondered at that oysters take alarm at this constant invasion of their domain and naturally seek some other place of rest ?"

Far more prejudicial to the welfare of the oysters than an occasional raid upon them by a stray Mutukurupam or Kallymuttu is, in all probability, the little mollusc, *súran*, which clusters in dense masses over large areas of the sea bottom, spreading over the surface of coral blocks, smothering and crowding out the recently deposited and delicate young of the oyster. Time after time there is, in the carefully kept records of the Superintendent of the Pearl Banks, in one year a note of the presence of young oysters, either pure or mixed with *súran* and mud or weed, while, at the next time of examination, generally in the following year, the oysters had disappeared, and the *súran* remained. A few examples will suffice to make this point clear :—

Devi Par [1]—$6\frac{1}{2}$ to $7\frac{1}{4}$ *fathoms.*

May 1881. Young oysters mixed with sooram [2] and mud.
,, 1882. Sooram.

Permandu Par—6 to $6\frac{1}{4}$ *fathoms.*

May 1880. A few oysters of one year age.
,, 1881. Young oysters mixed with sooram and mud.
,, 1882. Sooram.

Athombadu Par—$7\frac{3}{4}$ to 9 *fathoms.*

May 1880. Covered with sooram.
,, 1881. Large number of oysters of one year age, with sooram in some places and covered with weeds.
,, 1882. No oysters ; sooram in some places.

The bank, which was fished during the recent fishery, is situated about 10 miles east of Tuticorin, and known as the

---

[1] *Par* or *paar* = bank.   [2] *Sooram* = *súran*.

Tholayiram Par, the condition of which, as regards oyster supply, since the year 1860, is shown by the following extract from the records :—

April 1860. Plenty of oysters $3\frac{1}{2}$ years old.
Nov. 1861. Oysters scarce ; nearly all gone.
April 1863. Sooram and killikay with some young oysters.
Nov. 1865. ⎫
April 1866. ⎪
 ,, 1867. ⎬ Blank.
Nov. ,, ⎪
April 1869. ⎭
Mar. 1871. Five oysters with a quantity of sooram.
Feb. 1872. Five oysters of 3 years age found.
May 1873. Three oysters found.
Jan. 1875. Three oysters of 2 years age found.
Mar. 1876. North part blank.
April 1877. South part blank.
 ,, 1878. Thickly stocked with oysters of 1 year age.
May 1879. ⎫ Blank.
 ,, 1880. ⎭
 ,, 1881. Some oysters of 1 year mixed with killikay.
 ,, 1882. No living oysters; dead shells and sooram.
April 1883. Three oysters found.
Mar 1884. Plenty of oysters of one year age; clean and healthy.

From 1884 the bank was carefully watched, and the growth of the oysters continued steadily, unchecked by adverse conditions, as the following figures show :—

10 shells lifted.
⎧ March   1884 weighed   1   oz.
⎪ October   ,,   ,, $3\frac{3}{4}$,,
⎪ March   1885   ,, $6\frac{1}{4}$,,
⎪ October   ,,   ,,   7   ,,
⎨ April   1886   ,, $7\frac{1}{2}$,,
⎪ November   ,,   ,, $8\frac{1}{2}$,,
⎪ March   1887   ,, $10\frac{3}{4}$,,
⎪ October   ,,   ,,   13   ,,
⎩ November 1888   ,, $15\frac{1}{4}$,,

In November 1888 15,000 oysters were lifted and their product valued by expert pearl merchants at Rs. 206-13-9, *i.e.*, Rs. 13-12-8 per thousand [1] as shown by the following copy of the statement of valuation :—

---

[1] The product of 12,000 oysters lifted from the Ceylon pearl bank, the fishing of which took place synchronously with that of the Tuticorin bank, in November 1888 was valued at Rs. 122. A further sample of 12,650 oysters, lifted in February 1889, was valued at Rs. 142.

Description	Size in Basket	Number	Quantity in Chevu	Weight				Value	Total Value	Per Chevu	Per Kalungy
				Kaluṅgy	Manjady	Kaluṅgy	Manjady				
						Total					
						RS. A. P.	RS. A. P.				
Áni	20	1	158/320		$1\frac{3}{16}$		$1\frac{3}{16}$	43 3 0	43 3 0	25 star pagodas.	22 star pagodas.
Kuruvel	30	1	26/320		$\frac{9}{16}$		$\frac{3}{16}$	4 6 0	4 6 0	16 do.	10 do.
Kalippu	50	6	45/320		$1\frac{6}{16}$		$1\frac{6}{16}$	7 14 0	7 14 0	16 do.	5 do.
Pisal	50	3	..		$1\frac{3}{16}$		$1\frac{3}{16}$	0 10 0	0 10 0	4 do.	1 do.
Kodai	20	4	..		$1\frac{1}{8}$		$1\frac{1}{8}$	0 10 3	0 10 3	3 do.	..
Vadivu	100	..	176/320		$6\frac{1}{4}$		..	0 0 0	0 0 0	..	..
	200	..	112/320		7	1	..	77 0 0	77 0 0	..	..
	400	..	64/320		$6\frac{3}{4}$		$3\frac{1}{4}$	5 7 6	5 7 6	..	..
Do.	..	..	..		$3\frac{1}{4}$		8	7 0 5	7 0 5	..	..
Pisal	..	..	..		8		$1\frac{3}{16}$	0 0 0	0 0 0	..	7 do.
Kodai	..	..	..		$1\frac{3}{16}$		..				
Tul	600	..	..		..	$1\frac{3}{4}$	..	42 14 0	42 14 0	..	$3\frac{1}{2}$ do.
Mosio	800	..	..		..	$1\frac{9}{16}$	$1\frac{9}{16}$	16 6 0	16 6 0	..	..
Shell pearl	1,000	..	..		5	..	..	1 1 0	1 1 0	..	..
								Total ..	206 13 9		
								Average per 1,000 oysters ..	13 12 8		

It may not be out of place to elucidate the meaning of some of the terms used in the above statement, and I cannot do better than quote from the excellent article on the Pearl Fisheries of Ceylon by Mr. G. Vane, C.M.G., who writes as follows [1]:—

"Sorting and sizing the pearls into ten different sizes, from the largest to the smallest, is done by passing them through ten brass sieves of 20, 30, 50, 80, 100, 200, 400, 600, 800, and 1,000 holes ..... each of the ten sizes may include some of every class of pearls; the 20 to 80 and 100 may each have the *áṇi*, *anatari*, and *kallipú* kinds, and this necessitates the operation of classing, which requires great judgment on the part of the valuers.

"Perfection in pearls consists in shape and lustre, viz., sphericity and a silvery brightness, free from any discolouration; and, according as the pearls possess these essentials, the valuers assign their appropriate class, namely,—

"Áṇi	Perfect in sphericity and lustre.
"Anatari	Followers or companions, but failing somewhat in point of sphericity or lustre.
"Masaṇkú	Imperfect, failing in both points, especially in brilliancy of colour.
"Kaḷḷipú	Failing still more in both points.
"Kural	A double pearl, sometimes áṇi.
"Písal	Misshapen, clustered, more than two to each other.
"Maḍaṇku	Folded or bent pearls.
"Vaḍivu	Beauty of several sizes and classes.
"Túḷ	Small pearls of 800 to 1,000 size.

"The pearls having been thus sized and classed, each class is weighed and recorded in *kaḷañchu* (kalungy) and *mañcháḍi* (manjaday).

"The *kaḷañchu* is a brass weight equal, it is said, to 67 grains Troy. The *mañcháḍi* is a small red berry [2]; each berry, when full sized, is of nearly, or exactly the same weight; they are reckoned at twenty to the kaḷañchu.

"The weights being ascertained, the valuation is then fixed to each pearl class or set of pearls according to the respective sizes and classes: the inferior qualities solely according to weight in *kaḷañchu* and mañcháḍi; the superior *áñi*, *anatari*, and *vaḍivu* are not valued only by weight, but at so much *per chevo* of their weight, this *chevo* being the native or pearl valuer's mode of

---

[1] *Journal, Ceylon Branch, Royal Asiatic Society*, 1887, vol. X, No. 34. Paper read at the Conference Meeting of the Colonial and Indian Exhibition, October 6, 1886.
[2] The seed of *Abrus precatorius*.

assigning the proper value by weight to a valuable article of small weight, form and colour also considered."

The pearls of commerce are, of course, for the most part those which are formed within the soft tissues of the animal, and not the irregular pearly excrescences (*oddumutta*) which are found as outgrowths of the nacreous layer of the shell, frequently at the point of insertion of the adductor muscle. The nacreous layer of the Gulf of Manaar pearl oyster shell is very thin and of hardly any commercial value, the shells, after the extraction of the pearls by the process of decomposition, being used mainly in the manufacture of chunám.

As regards the cause of the formation of pearls, concerning which many theories have been hazarded, the most prevalent idea being that they are a "morbid secretion" produced as the result of disease, I may quote from the excellent "Guide to the Shell and Starfish Galleries in the British Museum (Natural History,)"[1] which tells us that some small foreign body, which has accidentally penetrated under the mantle and irritates the animal, is covered with successive concentric layers of nacre, thus attaining sometimes, but rarely, the size of a small filbert. The nacre is generally of the well-known pearly-white colour, very rarely dark, and occasionally almost black.[2] The effort of the animal to get rid of the irritation caused by a foreign substance between its valves, by covering it over with nacre, and thus converting it into a pearl, is strikingly illustrated by two specimens in which, in the one case, an entire fish, and, in the other, a small crab has been so enclosed. According to Streeter (*op. cit.*) the nucleus of the pearl may be either a grain of sand, the frustule of a diatom, a minute parasite, or one of the ova of the oysters, thin layers of carbonate of lime being deposited around the object concentrically, like the successive skins of an onion, until it is encysted.

Writing in 1859[3] as to what may be termed the worm theory of pearl formation, Dr. Kelaart stated that "as this report may fall into the hands of scientific men, I shall merely mention here that Monsieur Humbert, a Swiss Zoologist, has, by his own observations at the last pearl

---

[1] Printed by order of the Trustees, 1888.
[2] Among the pearls from the samples lifted at Tuticorin in November 1888 there is one dumb-bell shaped specimen of which one half is white, the other dark brown.
[3] *Report on the Natural History of the Pearl Oyster of Ceylon*, 1858-59.

fishery, corroborated all I have stated about the ovaria or genital glands and their contents, and that he has discovered, in addition to the filaria and cercaria, three other parasitical worms infesting the viscera and other parts of the pearl oyster. We both agree that these worms play an important part in the formation of pearls, and it may yet be found possible to infect pearls in other beds with these worms, and thus increase the quantity of these gems. The nucleus of an American pearl drawn by Möbius is nearly of the same form as the cercaria found in the pearl oysters of Ceylon." The " cercaria " referred to were, probably, Cestode worms (*Anthocephalus*, &c.), which are found in the internal organs of various fishes caught off the coast of Southern India, and gave rise to a scare in the European fish-loving community a few years ago. During the recent fishery in only a few out of many hundreds of oysters which I examined did I find small nemertine worms living on the mantle or gills of the oyster, so that their presence cannot be regarded as a common or essential occurrence.

The Gulf of Manaar pearl oyster (*Avicula fucata*, Gould) is represented in plate 1, as it appears after removal of one valve of its shell, the " ovarium," mantle, gills, adductor muscle, and byssus being exposed. The presence of a small pearl imbedded within the substance and projecting from the surface of the " ovarium " is indicated at A. The byssus (B), of which the function has given rise to much discussion and speculation, is made up of a bundle of tough, green-coloured fibres, secreted by a gland in the foot, and is capable of being protruded beyond, or retracted within the shell. By its means the animal is enabled to anchor itself on the sea-bottom to a neighbouring oyster or other mollusc shell, coral-block, melobesian nodule, or other convenient object; and it is said that the animal can, even in the adult stage, voluntarily shift its position and migrate to a considerable distance. That the young oyster can, during its phase of existence as a minute, free-swimming organism wander about and eventually settle down on some congenial spot no one will dispute; but the evidence that the adult oyster can, under natural conditions, migrate to any considerable distance is wholly insufficient, even though it has been demonstrated by experiments that a young pearl oyster under unnatural conditions in a soda-water tumbler full of sea-water can, though weighted with two other oysters of nearly its own size, climb up a smooth perpendicular surface

Plate I

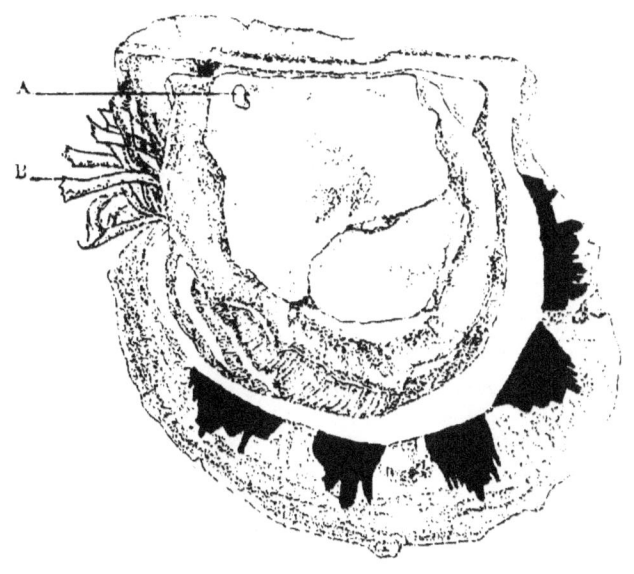

Avicula Fucata
Pearl Oyster

at the rate of an inch in two minutes. The mysterious disappearance of the oysters from the Ceylon pearl bank prior to a recent fishery must, I think, be attributed to the action of a strong under-current, and not to voluntary migration of the headless mollusc.

. . . .

The recent Tuticorin pearl fishery was carried on from a temporary improvised village, erected on the barren sandy shore at Salápatturai, 2 miles north of the town, and built out of palmyra and bamboo, the inflammability of which was demonstrated on more than one occasion. The village consisted of the divers' and merchants' quarters and bazárs, where, as the fishing progressed, the product of the oysters was exposed for sale; bungalows for the officials connected with the fishery; a tent used by myself as a zoological laboratory; dispensary; kottus (or koḍḍus), *i.e.*, enclosed spaces in which the counting, decomposition, and washing of the oysters are carried on; a Roman Catholic chapel; and the inevitable isolated cholera quarters.

The fishery commenced on the 25th of February under a combination of adverse conditions which seriously affected the revenue, viz., the presence of the pearl bank at a distance of 10 miles from the shore and in 10 fathoms of water, and the co-existence of a fishery on the Ceylon coast, where the oysters were to be obtained at a distance of about 5 miles from shore and at a depth of 5 to 7 fathoms. The natural result was that the natives, keenly alive to their own interests, went off with their boats from the Madras seaport towns of Pámban and Kílakarai to the Ceylon fishery, where they could earn their money more easily and with less discomfort than at Tuticorin, leaving the Tuticorin bank to be fished by a meagre fleet of about 40 boats.

An excellent account of the method of conducting the pearl fishery at Tuticorin has been published in the " Hand-Book of Directions to the Ports in the Presidency of Madras and Ceylon," 1878, from which the following varies only in points of detail.

The landwind, under favourable conditions, commences to blow soon after midnight, and a signal gun is fired by the beach master as a warning that the fleet of native boats, each with its complement of native divers, can start out to sea, their departure being accompanied by a good deal of noise and excitement. The bank should be reached by daylight, and the day's work commences on a signal being given from a schooner, which is moored on the bank throughout the fishery. An attempt is made to keep the boats together within an area marked out by buoys, so as to prevent the bank from being fished over in an irregular manner, and the temper of the European officer in charge of the schooner is sorely tried by the refusal of the boatmen to comply with the conditions. All being ready on board, a diving stone, weighing about 30 lbs., to which a rope is attached, and a basket or net fastened in a similar manner are placed over the ship's side. The ropes are grasped by the diver in his left hand, and, placing a foot on the stone, he draws a deep breath, and closes his nostrils with his right hand, or with a metal nose clip which he wears suspended round his neck by a string. At a given signal, the ropes are let go, and the diver soon reaches the bottom, his arrival there being indicated by the slackening of the rope. He then gets off the diving stone, which is drawn up to the surface, and, after filling the basket or net with oysters, if he is on a fertile spot, gives the rope a jerk, and comes up to the surface to regain his breath.

The contents of the basket or net are emptied into the boat, and the live oysters separated from the dead shells, débris, &c. The divers work in pairs, two to each stone, and the oysters which they bring up are kept separate from those of the other divers. A good diver will remain below the surface about 50 seconds, and, exceptionally, 60, 70 or even 90 seconds.

The largest number of oysters collected as the result of a single day's fishing by 41 boats during my visit to the fishery was 241,000, giving an average of 5,878 oysters per boat, a very small quantity when compared with the results of the Ceylon fishery in 1857, when the daily yield varied from one to one and a half million oysters, some boats bringing loads of thirty to forty thousand.

From experiments made with divers equipped with diving helmets, gathering stones instead of oysters, by the late Superintendent of the Madras Harbour Works, it was calcuculated [1] that a pair of helmeted divers could together send

---

[1] Vide *Madras Board of Revenue Resolution*, No. 677, dated 3rd August 1888.

up 12,000 shells an hour in shallow water, or, allowing for delay in hauling up in 12 fathoms of water, say, 9,000 shells an hour; and as, allowing for shifts, each diver should work four hours a day, the quantity sent up by a pair of divers in a day would be respectively $4 \times 12,000 = 48,000$, or $4 \times 9,000 = 36,000$ shells a day, which is equivalent to the work of 24 or 18 naked native divers sending up 2,000 a day.

The results of the work done by the two helmeted divers who were employed as an experiment at the Tuticorin fishery fell far short of this calculation, and compared unfavourably with the work done by the skilled native divers without helmets.

The diving operations cease for the day some time after noon, and the boats, if aided by a favourable sea breeze, reach the shore by 4 P.M., their arrival being awaited by large crowds of natives, some of whom come from curiosity, others to speculate on a small scale. On reaching the shore the boats are quickly made fast in the sand, and the oysters carried on the heads of the divers into the kottu, where they are divided into separate heaps, each set of divers dividing their day's produce into three equal portions. One of these, selected by the Superintendent of the Fishery or some other official, becomes the property of the divers, who quickly remove their share from the kottu, and, squatting on the sand, put their oysters up for sale at prices varying from about 15 to 40 for a rupee. On the first day of the fishery the oysters, for a short and to the divers lucrative time, were sold for four annas a piece. The two heaps which are left by the divers in the kottu become the property of Government, and are counted by coolies engaged for the purpose. Usually about 6 P.M. the Government oysters are sold by public auction, duly announced by tom-tom, being put up in lots of 1,000; and the purchaser can, subject to the consent of the auctioneer, take a certain number of thousands at the same rate as his winning bid. Occasionally a combination is organised among the merchants who are buying on a large scale, and come to the auction determined not to bid more than a very small fixed sum per 1,000. A struggle then takes place between the auctioneer and merchants, the former refusing to sell, the latter refusing to raise their price; and the struggle invariably ends in the collapse of the merchants when they find that their supply of oysters is cut off. No credit is allowed, and the buyers, as soon as they have paid their money into the treasury, remove their oysters to the washing kottus, or send them away up-country by railway.

Buyers of oysters on a very small scale open them at once with a knife, and extract the pearls by searching about in the flesh of the animal; but, by this method, a number of the very small pearls are missed, and it would be impossible to carry it out when dealing with oysters in large numbers. Boiling the oysters in water and subsequent extraction of the pearls from the dried residue might be, with advantage, resorted to as a more wholesome and less unsavoury process than the one which is commonly resorted to of leaving the oysters to putrify in the sun, and subsequently extracting the pearls from the residue after it has been submitted to repeated washings to free it from the prevailing maggots, pulpy animal matter, sand, &c. The process of putrefaction is greatly aided by flies—big red-eyed blue-bottles. At the Ceylon pearl fishery, which I was sent to inspect on the termination of my work at Tuticorin, the merchants complained at first of the scarcity of flies; but, later on, there was no cause for complaint, for they were present not only in the kottus, but in other parts of the camp, in such enormous numbers as to form a veritable plague, covering our clothes with a thick black mass, and rendering the taking of food and drink a difficult and unpleasant process until the evening, when they went to rest after twelve hours of unceasing activity.

For months after the conclusion of a pearl fishery poor natives may be seen hunting in the sand on the site of the pearl camp for pearls, and it is reported that in 1797 a common fellow, of the lowest class, thus got by accident the most valuable pearl seen that season, and sold it for a large sum.

. . . . . .

Towards the latter end of 1888 it was suggested that an electric light apparatus should be acquired in connection with the pearl fishery, by means of which one would be able to examine the condition of the bank from the deck of a ship, and which, it was thought, would help to solve the enigmas that still hang about the migrations of the pearl oyster. The notice of Government was drawn to the fact that a boat had been fitted up with a brush-dynamo and electric globe for the pearl fishery in South Australia by a Glasgow firm. During a recent visit to Europe, I made a series of inquiries as to the possibility of obtaining a light, such as was required; but, though there was abundant evidence as to the use of the electric light for surface work, salvage operations, and scientific dredging,[1] the general opinion of those best qualified to judge was that it would, for the proposed purpose, be a failure. It has been suggested by Mr. G. W. Phipps, who was for many years Superintendent of the Tuticorin pearl banks, that, if a sheet of thick glass could be let into the lower plates of a vessel and there protected both outside and inside in some way from accident, a study of the sea-bottom in clear water, either by day with the sun's rays or by night by the use of a powerful electric light, might be made. In a letter to Government Mr. C. E. Fryer, Inspector of Fisheries, makes the sound suggestion "that the observations which the Government of Madras desire to make upon the habits of the pearl oysters would be greatly facilitated by the employment of a diver equipped with an ordinary diving dress. By this means a prolonged stay could be made by an observer on the sea-bottom, who could not only make an accurate survey of the bed, but could periodically examine the same ground, select specimens, and make minute observations, which would be impossible to a native diver, whose stay at the bottom is limited to a minute or so." To these remarks I may add my own experience at the Tuticorin fishery, where, by examination of the shells of the oysters brought up by the divers, by expending small sums of money which tempted the native divers to bring me such marine animals as they met with at the sea-bottom, by conversation with the European diver, who was, further, able to bring up large coral blocks (*Porites, Madrepora, Hydnophora, Pocillopora, Turbinaria*, &c.) for examination, and by dredging, I was able to form some idea as to the conditions under which the pearl oysters were living. On clear days it was possible to distinguish the sandy from the rocky patches by the effect of light and shade, and from hauls of the dredge over the former not only many mollusca, &c., but also specimens

---

[1] *Vide* Herdman's 2nd *Annual Report on the Puffin Island Biological Station.*

of *Branchiostoma*, sp.[1] (Lancelet) were obtained, of which the largest measured two inches in length. Mollusca were also obtained in great variety by passing the débris, which was swept from the floor of the kottoo every day after the oysters had been cleared away, through sieves. The big *Murex anguliferus* (Elephant Chank) was brought in from the banks by the divers nearly every day, and the animal served up for their hard-earned evening meal. The oysters shells were largely encrusted with bright-coloured sponges, of which the most conspicuous was *Clathria indica* (n. sp.) an erect-growing bright red species, recorded by Mr. Dendy in his report on my second collection of sponges from the Gulf of Manaar.[2] Very abundant, too, was the large cup-shaped *Petrosia testudinaria*, of which a specimen in the Madras Museum measures 1·5 feet in height. Enveloping the oyster shells were tangled masses of marine *Algæ*,[3] and floating in dense masses on the surface was the Sargasso weed, *Sargassum vulgare*. The various minute living organisms entangled in the meshes of the Algæ must serve as an efficient food-supply for the oysters. The outer surface of the living oyster shells was frequently covered with delicate *Polyzoa*, which also flourished on the internal surface of the dead shells in the form of flat or arborescent colonies. In no single instance did I see an oyster shell from the Tuticorin bank encrusted with coral; whereas at the Ceylon fishery, on the sole occasion on which I had an opportunity of examining the oysters brought in from the pearl bank, I found the surface of a large number of the shells, both dead and living, covered, and frequently entirely hidden from view by delicate branching *Madrepora* or *Pocillopora*, or the more massive *Astræa*, *Cœloria*, *Hydnophora*, *Galaxea*, &c. A specimen of *Galaxea* encrusting a single valve of an oyster shell, which I picked up on the shore and is now in the Madras Museum, weighed as much as 5 oz. 15 dwts.

Several species of *Echinoderm*, which have not hitherto been recorded from the coast of the Madras Presidency,[4]

---

[1] Specimens of *Amphioxus belcheri*, Gray, were obtained by Mr. Giles, when dredging from the Marine Survey SS. "Investigator" off Seven Pagodas (Mahábalipuram) 30 miles south of Madras during the season 1887-88.
[2] *Ann. Mag.*, *Nat. Hist.*, Feb. 1889.
[3] The collection of Algæ made at Tuticorin has been sent to the British Museum (Nat. History) for identification.
[4] Vide *Proc., Zool. Soc., Lond.*, June 19, 1888.

were brought up by the divers, and have been sent to my friend Professor Jeffrey Bell for identification. Of recorded species those which were brought on shore most frequently were the crimson-lake coloured *Oreaster lincki*, and the long-armed, usually salmon-coloured *Linckia lævigata*, and, not unfrequently, dense clusters of *Antedon palmata* were found in crevices hollowed out in coral blocks, from which also, when broken open, specimens of *Ophiuroids* (commonly met with their arms turned round the branches of a *Gorgonia*, or in the canal system of sponges), *Annelids, Crustaceans,* and stone-boring *Mollusca* (*Lithodomus, Parapholas,* *Venerupis,* &c.) were obtained.

II.—NOTE ON PEARLS FROM MYTILUS AND PLACUNA.

II.—NOTE ON PEARLS FROM MYTILUS AND PLACUNA.

In addition to the pearl oyster of the Gulf of Mannar, two other pearl-producing mollusca (*Mytilus smaragdinus* and *Placuna placenta*) are to be found in the Madras Presidency: the former in the Sonnapore river in the Ganjam district, where they are, or were till recently, the source of a local industry; the latter on salt mud flats and in canals in various parts of the presidency, *e.g.*, Pulicat Lake, the Buckingham Canal, Tuticorin, &c.

As regards the former (*M. smaragdinus*), samples of the pearls were sent to Government by Mr. R. Davison, when Acting Collector of Ganjam in 1875, and examined by pearl merchants, who reported that they were of very inferior quality and of the description termed "rejected pearls" by the trade, and valued a big discoloured pearl at Rs. 1-8-0 and the whole sample at Rs. 7. The following extract is taken from a letter to Government by Mr. Davison, who, as the result of a visit to the mussel beds, which was resented by the natives who were interested in keeping the habitat of the mussels secret, suggested that, if taken in hand and properly treated, the pearls might eventually become a fruitful source of revenue :—

"Sonnapore is a small fishing village situated near the mouth of a river to which it gives its name, and which is about 12 miles south of Gopaulpore. For some miles up the river there are large beds of the ordinary edible oysters, which find a ready market at Berhampore and elsewhere. Mixed up with the ordinary oysters, and adhering most tenaciously to their beds, are the bright green mussels, from which the pearls are produced.

"I had five canoes, with four divers in each, at work, and the place where we were most successful is situated about two miles from the mouth of the river and about half a mile beyond the custom house. Each diver brought with him a long bamboo pole, which he drove with all his might into the oyster bed at a depth of from ten to twelve feet of water according to the state of the tide. He then dived to the bottom, and holding on to, or keeping near the bamboo, broke off as large a mass of oysters as he could conveniently bring to the surface in one hand, and with the other he helped himself up the bamboo. Any mussels that were found adhering to the block of oysters were secured, and the oysters were returned to the water, as thousands of them have already, from time to time, been examined in vain. I was amazed at the dexterity and rapidity with which the divers opened the mussels with knives made for the purpose; and the expert manner in which they ran their thumbs over the molluscs, detecting in an instant without fail the most minute seed pearl not larger than a pin's head, leaves no room for doubting that long practice has made them perfect in this particular branch of what has hitherto been to them a highly lucrative employment."

The flat, transparent shells of *Placuna placenta* (window shell) are used in China and at the Indo-Portuguese Settlement Goa as a substitute for window glass; and the small pearls which the animal produces are exported to India to be calcined into chunám, which rich natives chew with their well-beloved betel, and are said to be burned in the mouths of the dead.[1] So far as I am aware the pearls which might be obtained from the masses of *Placuna* which live in the mud flats of Southern India have not been utilised as an article of commerce. But an extensive *Placuna* pearl fishery has been carried on at Tamblegam lake in Ceylon; and some idea of the abundance of the mollusc may be gathered from the fact that the quantity of shells taken in the three years prior to 1858 could not have been less than eighteen millions.

---

[1] *Vide* Tennent's *Ceylon*, II, 492.

III.—THE TUTICORIN CHANK FISHERY.

Plate II

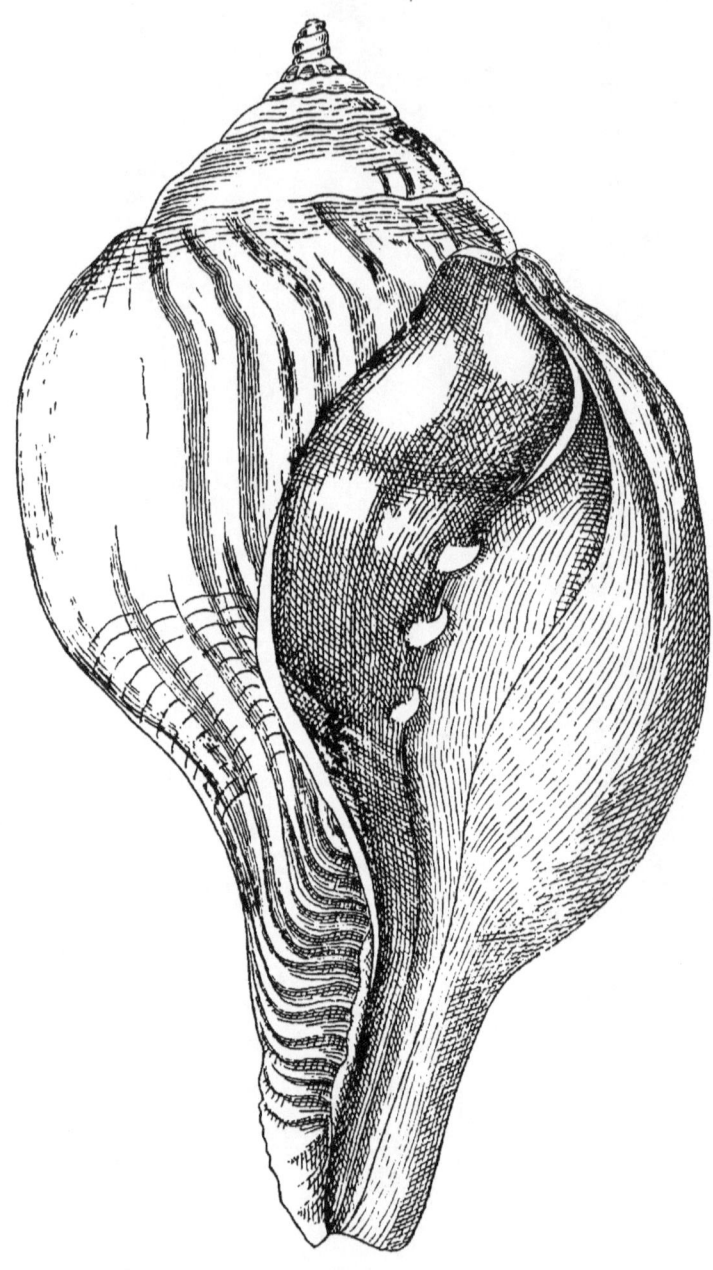

TURBINELLA RAPA
THE CHANK

III.—THE TUTICORIN CHANK FISHERY.

The sacred chank or *sankh* is the shell of the gastropod mollusc *Turbinella rapa*, of which a full-grown specimen is represented on plate II, and is, like the pearl oyster and the edible trepang (*Holothuria marmorata*) a commercial product of the Gulf of Manaar.

The chank which one sees suspended on the forehead and round the necks of bullocks in Madras is not only used by Hindus for offering libations and as a musical instrument in temples, but is also cut into armlets, bracelets, and other ornaments, and writing in the sixteenth century Garcia says [1]:—

" And this *chanco* is a ware for the Bengal trade, and formerly produced more profit than now . . . . . . and there was formerly a custom in Bengal that no virgin in honour and esteem could be corrupted unless it were by placing bracelets of *chanco* on her arms; but since the Patáns came in, this usage has more or less ceased, and so the *chanco* is rated lower now."

The chank appears as a symbol on some of the coins of the Chalukyan and Pándyan empires, and on the modern coins of the Rájas of Travancore.

The chank fishery is conducted from Tuticorin, and the shells are found in the vicinity of the pearl banks, in about 7 to 10 fathoms,[2] either buried in the sand, lying on the sea bottom, or in sandy crevices between blocks of coral rock. The fishery goes on during the north-east monsoon from October to May, and is worked by native divers, who, putting their foot on a stone to which a long rope is attached, are let down to the bottom, carrying a net round the waist, in which they place the chanks as they collect them. The shells of the chank are scattered about, and not aggregated together in clusters like those of the pearl oyster, so that the

---

[1] *Vide* Yule and Burnell *Hobson Jobson*, 1886.
[2] For a discussion of the chank as an enemy of the pearl oyster, *vide* Mr. H. S. Thomas' *Report on Pearl Fisheries and Chank Fisheries*, 1884.

divers have to move about on the bottom from place to place in search of them. The divers usually stay beneath the surface from 40 to 50 seconds. The longest dive which I have myself witnessed was 54 seconds, and on that occasion the diver, on his return to the surface, innocently inquired how many minutes he had been under water. A single case is on record of a native diver being drowned from overloading his net, so that he was unable to rise to the surface. I can find no record of the death, in recent years, of a diver at the hands of a shark; but dread of sharks still clings to the divers, and I read in the *Times of Ceylon* during the recent pearl fishery that " at present there are said to be 150 boats with their full complement of men, all waiting at Kilakarai in readiness to proceed to Dutch Bay, but they will not leave until after some festivities which occur on the 15th instant, when it is customary for them to pray for protection from sharks, &c., while engaged in diving."

Further Tennent writes [1] :—

"The only precaution to which the Ceylon diver devotedly resorts is the mystic ceremony of the shark-charmer, whose exorcism is an indispensable preliminary to every fishery. This power is believed to be hereditary; nor is it supposed that the value of his incantations is at all dependent upon the religious faith professed by the operator, for the present head of the family happens to be a Roman Catholic. At the time of our visit this mysterious functionary was ill and unable to attend; but he sent an accredited substitute, who assured me that, although he was himself ignorant of the grand and mystic secret, the fact of his presence as a representative of the higher authority would be recognised and respected by the sharks."

The number of chanks collected in a day varies very much according to the number of divers employed and other conditions; and the records show that as many as 6,000 or as few as 400 may be collected. The divers, who are furnished with canoes, ropes, and other apparatus, are paid at the rate of Rs. 20 per 1,000 shells. At the close of the day's fishery the chanks are brought on shore, and examined. Those which are defective, either from cracks or irregularities of the surface from their having been gnawed by fishes or bored by marine worms, are rejected.

The remainder are tested with a wooden gauge having a hole $2\frac{3}{8}$ inches in diameter. Those shells which pass through

---

[1] Sir J. Emerson Tennent's *Ceylon*, 1860, vol. II, pp. 564-5.

this hole are discarded as being too small, and returned to the sea on the chance that the animal may revive and continue to grow; those which are too large to pass through the hole are stored in a "godown," where the animal substance is got rid of by the process of putrifaction, assisted by flies and other insects. In the month of July the shells are sold by auction in one lot to the highest bidder. In 1886 the highest offer was Rs. 96 per 1,000 by a native of Kílakarai, which was accepted.

The following statement shows the number of chank shells fished, and the net amount realised from 1876-77 to 1885-86:—

Years.	Chanks fished.	Net amount realised.
		RS. A. P.
1876-77	282,737	12,066 4 6
1877-78	360,131	22,904 4 7
1878-79	388,064	22,250 4 7
1879-80	125,540	6,714 13 2
1880-81	105,277	9,645 8 3
1881-82	303,590	28,450 8 6
1882-83	247,696	22,038 13 7
1883-84	210,005	11,347 1 5
1884-85	..	....
1885-86	332,757	23,970 0 11
Total ..	..	1,59,387 11 6

It would seem from Simmond's "Commercial Products of the Sea" that the chank fishery was, in days gone by, more lucrative than it is at present; for it is there stated that "frequently 4,000,000 or 5,000,000 of these shells are shipped in a year from the Gulf of Manaar. In some years the value of the rough shells, as imported into Madras and Calcutta, reaches a value of £10,000 or £15,000. The chank fishery at Ceylon at one time employed 600 divers, and yielded a revenue to the Island Government of £4,000 per annum for licenses."

A right-handed chank (i.e., one which has its spiral opening to the right), which was found off the coast of Ceylon at Jaffna in 1887, was sold for Rs. 700. Such a chank is said to have been sometimes priced at a lakh of rupees; and, writing in 1813, Milburn says[1] that a chank

---

[1] *Oriental Commerce*, vol. I, p. 357.

opening to the right hand is highly valued, and always sells for its weight in gold. Further, Baldæus, writing towards the end of the seventeenth century, narrates the legend that " Garroude flew in all haste to Brahma and brought to Kistna the *chianko* or kinkhorn twisted to the right."

The curious egg capsules of the chank, of which many specimens were brought up for me by the Tuticorin divers, have been well described by my predecessor, Dr. G. Bidie, C.I.E., who says of them [1]:—

" The spawn of the Turbinella consists of a series of sacs or oviferous receptacles (pl. iii, fig. 1), the transverse markings in the figure indicating the dimensions of each capsule. In the fresh state the membranous walls of the sacs are pliable, although tough and horny; and it will be observed that, during the drying process, the spawn has, from the irregular shrinking of the two sides, become curved and twisted so as to have somewhat the appearance of a horn . . . . . In fig. 2 a side view is given of a separated capsule, and fig. 3 gives magnified sketches of the young shells. The larger oviferous sacs of the Turbinella spawn contain from 8 to 10 young shells each, but the smaller ones, towards the end of the specimen, are barren. In fig. 1 there are 30 fertile sacs, and, say that each of these on an average contains 6 germs, we thus have altogether 180 young shells in the whole of the cells."

The largest number of young shells which I found in a single specimen was 235, of which the average diameter was ·62 inch.

---

[1] *Madras Journal of Literature and Science*, vol. XXIV, 1879, pp. 232-234.

Plate III

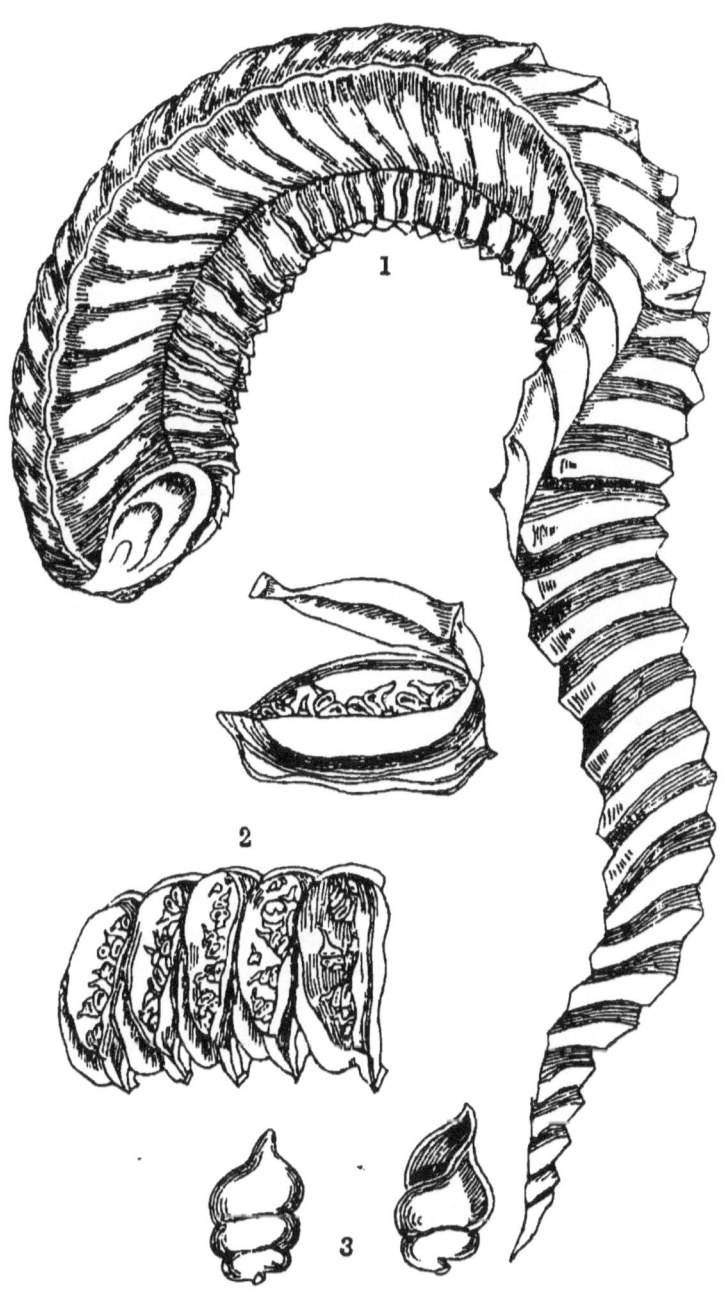

EGG CAPSULE OF CHANK

IV.—NOTE ON THE CEYLON PEARL FISHERY, 1889.

IV.—NOTE ON THE CEYLON PEARL FISHERY, 1889.

On the completion of my investigation of the Tuticorin pearl fishery, in accordance with instructions received from Government, I proceeded to Ceylon to report on the pearl fishery which was being carried out at Dutch Bay.

It was originally intended that I should travel up the coast by S.S. *Active;* but, as she was laden with stores for the pearl camp, there was no available space, and I had, unfortunately, to wait for a passage on the small coasting steamer *Prince Alfred*, which left Colombo two days later. As we neared Dutch Bay in the early morning, the well known odour of decomposing oysters was perceptible some distance out at sea, and we watched nine boats at work on the pearl bank. A single haul of the dredge in the shallow water of the bay brought up a number of small mollusca, worms, and a gephyrean, which I had not seen before. Arrived at the camp I found Mr. Twynham, C.M.G., Captain Donnan, (whose name is connected with a Gulf of Manaar sponge *Axinella donnani*), and other administrative officers living on board S.S. *Serendib* which was moored close to the shore, communication with which was maintained by means of a gangway. Several deaths from cholera occurred on board during the return journey of the *Serendib* to Colombo, and, among others Captain Robson, who had acted as Kottu Superintendent throughout the fishery, fell a victim to the dread disease.

The nine boats which had been at work on the bank were towed into the bay by the *Active*, reaching the shore opposite the kottus before 4 p.m. I gathered that the steamer had been of very great service during the fishery; for, with her assistance, not only were the boats enabled to get to and from the bank in spite of contrary winds, but the work of the divers, which is very severe, was considerably lightened by the simple fact that the steamer could bring them back at an early hour on days when, without her assistance, they would have been out at sea until late in the evening and not ready to start off for the bank on the following morning.

Fortunately I examined the oysters which were brought in by the boats; for, as events turned out, it was my only opportunity of making an examination. I was at once struck with the fact that the shells of the oysters presented an entirely different appearance to those of the Tholayiram Par (Tuticorin). For, whereas the latter were enveloped in dense masses of algæ (sea weeds) and the surface of the shells was covered by variously coloured branching and sessile encrusting sponges, the surface of the shells of the former which was uppermost during life was, in very many cases, covered over by stony corals, which formed either encrusting masses or branching tufts. Specimens of the shells, with their accompanying corals, many of which were to be seen lying strewn along the sandy shores of the bay, discarded by natives after extraction of their contents, have been deposited in the Madras Museum. Further examination of these coral-bearing shells would be of interest; for, as the age of the oysters can be approximately fixed, a very good idea could be obtained, by weighing and by observation of the size of the corals on oysters of different ages, as to the rate at which the corals grow.[1] Chemical analyses of the sea water over the Ceylon and Tuticorin pearl banks, especially with reference to the percentage of lime salts, should also be carried out.

As regards my observation that the Tuticorin shells were covered with algæ while the Ceylon shells were encrusted by corals a Ceylon correspondent writes as follows:—

"From the fishery of 1887 we took away specimens, very beautiful to look at, but several of which showed that the unfortunate animals inhabiting the shells had their residences converted into their tombs by the fatal industry of the coral animals. But our specimens were not obtained from the Modaragam Par, which was that we saw fished, and the shells taken from which are always covered with red colored algæ, and never with corals. We gathered our coral-covered specimens from the mounds of dried shells on the sea-shore, and learned that they had been taken in a previous fishery from another bank."

The mid-day heat at Dutch Bay was very intense, the sand getting so hot that even coolies could not walk on it, and the blue-bottle flies were an intolerable nuisance from early morn till sun down. The plague of flies at the Ceylon

---

[1] The rate of growth of corals is fully discussed in Darwin's *Structure and Distribution of Coral Reefs*, 3rd ed., 1889.

fisheries has occurred on former occasions, and Mr. G. Vane, who conducted the fisheries from 1855-60, writes as follows :

"Then come flies, innumerable, of the largest kind ; indeed flies are constant plagues, but are worse with a southerly wind, everything being covered with a black mass; a glass of wine or water must be drunk as poured out, or it is filled with flies, but southerly winds do not last long, and it seems as though providentially arranged that the prevailing winds should aid the purposes and needs of a pearl fishery."

Early in the morning of the day following my arrival at Dutch Bay my suspicion that something was wrong was confirmed by the receipt of information that deaths from cholera had occurred in camp, and that there was a panic among the divers, who had struck work. It was promptly decided to abandon the fishery, and permission was given for the boats to leave. The divers' quarters and sale kottus, the fences of which had begun to throw out leaves, were, as a matter of precaution, burned down, and by 4 P.M. most of the boats were out at sea, many making for the Madras coast.

The general arrangement of the Dutch Bay camp corresponded, in all essential particulars, with the arrangement of the Tuticorin camp. The latter is, in fact, based on what I may term the Ceylon type.

The camp is described by a newspaper correspondent in the following words [1]:—

" What was only the other day a sandy desert is now a populous and thriving town, with rows of buildings and well-planned streets. The two principal streets run parallel to each other. Each is about a mile long and 120 feet wide. These are again intersected by cross roads at intervals of 200 feet, an arrangement which permits of free ventilation, &c. Along the centre of each principal street there is a row of wells and lamps .... That portion of the town described above is situated at the south end of Dutch Bay, and is occupied by merchants, boutique-keepers, divers, *et hoc genus omne*. To the west of this, where the buildings are of a superior order and more apart from each other, we have the custom-house, court-house, police station with the Union Jack flying gaily in front of it, the Government Auditor's quarters, the doctor's buildings, the general hospital, out-door dispensary, rest-house, &c. On the spit of sand (a sand bank) are built the Government and private kottus and the sale

---

[1] *Ceylon Observer*, 2nd March 1889.

bungalow. Here, too, are the headquarters of the police.... By the side of this spit of land, and closely moored to it, are the *Dib*, the *Antelope*, and the *Sultan Iskander*, which serve as quarters of the *Government Auditor, Captain Donnan*, and their subordinate officers. Far away from this site and at the very end of the spit can be described some half a dozen yellow flags, which are said to indicate the situation of the quarantine station and the hospitals for cholera and small-pox patients..... Somewhere about the commencement of the spit stands a dilapidated Roman Catholic church, sea-eaten and falling into ruins. Father Dineaux, who is temporarily in charge, tells me that his church is in imminent danger of total disappearance owing to encroachments from the sea like the proverbial building that was built on the sands. The cemetery which belonged to this church and formed part of its grounds has long since been claimed by the sea, and those who were once buried in terra firma now sleep beneath the wave."

A small guard steamer was employed in cruising about the bay during the fishery, so as to prevent the divers, on their return from the bank, from dropping bags of oysters in the shallow water, which could afterwards be picked up. This form of fraud—and the frauds perpetrated by pearl divers are many—was scarcely possible at Tuticorin, where the boats arrived on shore opposite the kottu straight from the open sea.

Good fresh water was obtained from shallow wells dug in the sandy shore, and there was an abundance of water, condensed by the *Serendib*, in a large tank; but the condensed water did not seem to be appreciated by the natives.

I had, unfortunately, no opportunity of watching the process of counting the oysters in the kottu, or the management of an auction on a large scale; but, so far as I could gather from the counting and sale of the oysters brought in by the nine boats already referred to, the system was the same as that adopted at Tuticorin.

Turning now to a comparison of the Tuticorin and Dutch Bay fisheries in the present year, the latter had the advantages of—

    i. a large fleet (193) of boats, and a correspondingly large staff of divers;
    ii. the presence of an efficient steam-tug throughout the fishery, by means of which both time and labour were saved;

iii. the existence of the oysters in comparatively shallow water and near to land.

The Tuticorin fishery laboured, on the other hand, under the disadvantages of—

    i. a very small fleet (44) of boats, and small staff of divers.

    ii. the absence of a tug for a long time after the commencement of the fishery.

    iii. the existence of the oysters in deeper water, and at a great distance from the shore than at Dutch Bay.

And there was, if the health of the camp is left out of the question, no compensatory advantage at Tuticorin.

The following table shows the results of the Dutch Bay fishery from the date of its commencement up to March the 27th :—

Date.	Number of boats.	Total number of oysters fished.	Sold for Government.	Average rate per 1,000.	Revenue.
				RS.	RS.
2nd March	89	542,527	361,685	28	10,133·87
4th ,,	170	1,030,342	686,895	22	14,340·80
5th ,,	..	1,183,455	788,970	28·79	22,718·10
6th ,,	191	1,343,415	895,610	26·19	23,461·47
7th ,,	188	1,611,616	1,074,410	20·00	21,488·20
8th ,,	..	1,357,365	904,910	20·05	18,143·11
9th ,,	190	1,432,717	955,145	21·96	20,983·19
11th ,,	193	1,623,750	1,082,500	20·17	21,834·00
12th ,,	191	1,688,430	1,125,620	15·01	16,909·30
13th ,,	190	1,599,045	1,066,030	15·00	15,990·45
14th ,,	190	1,803,240	1,202,160	16·44	19,769·56
15th ,,	187	1,926,000	1,284,000	19·04	24,453·00
16th ,,	190	2,209,688	1,473,125	21·63	31,868·75
18th ,,	191	1,992,847	1,328,565	19·31	25,656·30
19th ,,	189	2,439,802	1,626,535	15·95	25,956·03
20th ,,	188	1,946,250	1,297,500	15·00	19,462·50
21st ,,	190	2,238,998	1,492,665	19·95	29,781·63
22nd ,,	189	2,215,725	1,477,150	22·55	33,320·15
23rd ,,	187	2,372,003	1,581,335	18·36	29,035·70
25th ,,	187	..	1,325,875	15	19,888·13
26th ,,	..	..	1,099,070	17	17,730·12
27th ,, -	..	..	1,052,045	17	18,913·86

The total quantity of the Government share of oysters, was, therefore, 25,184,015, and the total sum realised as the result of 22 days' fishing Rs. 4,81,887·52.

Comparing these results with those of the Tuticorin fishery, the following table shows the results obtained at the latter during the time of the Dutch Bay fishery, viz., from 2nd March to 27th March :—

Date.	Number of boats.	Total number of oysters.	European diver.	Bombay diver.	Sold for Government.	Rate per 1,000.			Revenue.		
						RS.	A.	P.	RS.	A.	P.
2nd March ...	3	6,000	...	...	4,000	43	0	0	172	0	0
4th ,, ...	...	...	...	...	...	...			...		
5th ,, ...	38	151,500	...	...	101,000	25	6	4	2,565	0	0
6th ,, ...	38	180,000	...	...	120,000	25	13	2	3,099	0	0
7th ,, ...	40	180,000	...	...	120,000	24	14	3	2,987	0	0
8th ,, ...	41	187,333	254	80	125,000	26	1	5	3,261	0	0
9th ,, ...	42	224,654	130	562	150,000	25	6	8	3,813	0	0
11th ,, ...	44	204,907	592	594	137,000	22	10	3	3,102	0	0
12th ,, ...	42	235,121	643	115	157,000	21	0	4	3,301	0	0
13th ,, ...	44	235.017	1,405	760	158,000	21	3	2	3,350	0	0
14th ,, ...	37	148,280	439	...	99,000	21	8	5	2,131	0	0
15th ,, ...	35	158,905	190	...	106,000	20	10	8	2,191	0	0
16th ,, ...	44	213,809	2,000	2,381	144,000	21	2	6	3,067	0	0
18th ,, ...	...	...	...	...	...	...			...		
19th ,, ...	24	97,450	99	...	65,000	26	10	1	1,731	0	0
20th ,, ...	12	82,500	...	...	55,000	26	13	4	1,476	0	0
21st ,, ...	43	360,572	966	890	241,000	22	2	7	5,341	0	0
22nd ,, ...	44	292,473	1,452	1,602	196,000	21	12	9	4,274	0	0
23rd ,, ...	35	244,500	...	...	163,000	22	5	7	3,643	0	0
25th ,, ...	...	...	...	...	...	...			...		
26th ,, ...	2	4,565	3,070	1,000	4,400	30	5	0	133	6	5
27th ,, ...	44	379,025	950	...	253,000	24	10	2	6,234	0	0

The total quantity of the Government share of oysters, was, therefore, 2,398,400, and the total sum realised during the time under notice Rs. 55,871-6-5.

A comparison of these two tables is very instructive and brings out very clearly the fact that, whereas at Dutch Bay the fishery was carried on without interruption (no fishery took place either at Dutch Bay or Tuticorin on Sunday the 3rd, 10th, 17th and 24th), and, after the first few days, during which time all the boats had not arrived, or were not ready for work, a large and uniform number of boats were at work daily and regularly bringing in good loads of oysters; at Tuticorin, on the other hand, not only was there no fishery at all on three days (exclusive of Sundays), but on different occasions, out of the entire fleet of 44 boats, as few as 2, 3, and 12 boats were at work, with the result that, during 6 out of the 22 working days under review, only 63,400 oysters, yielding Rs. 1,781-6-5, fell to the Government share, i.e., the total yield of six days was less than that which was, with one exception, the 19th, obtained as the result of a single day's work.

V.—RÁMÉSVARAM ISLAND.

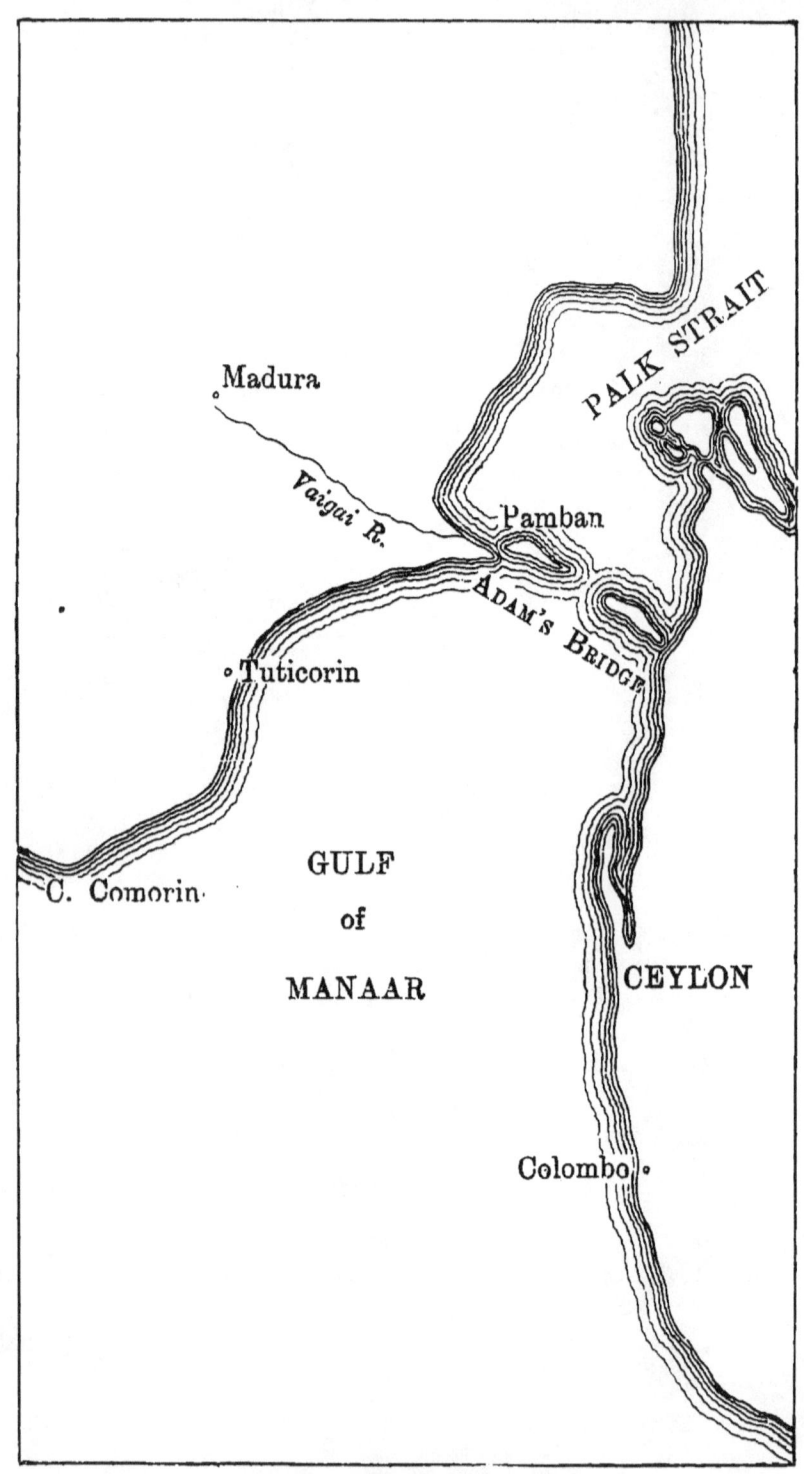

GULF OF MANAAR

V.—RÁMÉSVARAM ISLAND.

In January 1887 it was my privilege to accompany the Secretary to Government, Public Works Department, and the Presidency Port Officer, on a tour of inspection of lighthouses, from Mangalore on the west to Gopálpur on the east, which come within the jurisdiction of the Madras Government. My knowledge of the littoral of the Madras Presidency was, apart from Madras, at that time confined to Rámésvaram Island on which I had spent a few days in 1886, and the West Coast from Cochin to Trivandrum which I had visited, with a view to making a collection of the fishes of Malabar, soon after my arrival in India in 1885. The tour of light-house inspection afforded me an excellent opportunity, even though the halts at the light-house stations were, as a rule, very short, of forming a general idea as to the zoological capacity of the different parts of the coast, and deciding, by a visit to the fish bazárs and cursory examination of ' specimens ' cast up on shore, which afford, in some measure, an index to the still living and submerged fauna of the neighbouring sea, what parts of the coast were likely to afford the most profitable field for future investigation.

A casual non-scientific observer, walking along the sandy surf-beaten beach at Madras, will probably notice nothing to attract his attention except a number of coarse shells destined for the manufacture of chunám, an occasional flattened jelly-fish, and swift-footed crabs (*Ocypoda*) which, on the approach of man, scamper away, and disappear, like rabbits, within their burrows. But, if the same observer walks along the shore at Pámban or Hare Island, Tuticorin, he cannot help observing that it is strewn with broken fragments of dead coral, among which branches of *Madrepores* are most conspicuous, and sponges brought on shore by a recent tide, or dried up above tide-mark. And, if he trusts himself upon the slimy corals which are exposed at low tide, and turns them over so as to display their under-surface, he will find hidden there a wealth of marine animals—crabs, boring anemones, worms, shell-fish, *béches-de-mer*, and bright-coloured encrusting sponges. And the Madras beach may, allowing for differences of species, be taken as fairly representative of the coast of the Presidency with the exception of the coral-fringed shores of the islands which skirt the coast of the Gulf of Mannar, which I have visited on several occasions in the months of July and August. These months, though warm, proved very favorable, owing to the absence of rain, for carrying out investigations, and for the drying of specimens, *e.g.*, stuffed fishes, big sponges, and corals, such as are not suitable for preservation in alcohol or other fluid medium. Even, however, under the most favorable climatic conditions, the work of a marine zoologist beneath a tropical sun is, apart from the personal discomfort caused by the sun and glare on the water, except in the early morning and towards sunset, attended by many difficulties, which are graphically described by Haeckel, who says,[1] speaking of surface-netting with a gauze tow-net :—

"The wealth of varieties of marine creatures to be found in the Bay of Belligam was evident even on my first expedition. The glass vessels, into which I turned the floating inhabitants of the ocean out of the gauze net, were quite full in a few hours. Elegant *Medusæ*, and beautiful *Siphonophora* were swimming among thousands of little crabs and *Salpæ*; numbers of larvæ of mollusca were rushing about, mingled with fluttering *Hyaleadæ* and other pteropoda, while swarms of the larvæ of worms, crustacea, and corals, fell a helpless prey to greedy *Sagittæ*. Almost all the creatures are colourless, and as perfectly transparent as the sea-water in which they carry on their hard struggle for existence, which, indeed, on the Darwinian principle of selection, has given rise to the transparency of these pelagic creatures. But I soon discovered to my grief that, within a very short time after being captured, at most half an hour and often not more than a quarter, most of the fragile creatures died; their hyaline bodies grew more opaque, and, even before we could reach the land, I perceived the characteristic odour exhaled by the soft and rapidly decomposing bodies."

Haeckel's experience is, unfortunately, not an uncommon one, and, while staying at Pámban, I frequently had the mortification of finding, on my return from a surface-netting expedition to my improvised laboratory at the zemindary bungalow, instead of a crowd of living creatures,

---

[1] *Visit to Ceylon.* Transl. by Clara Bell, 1883.

an amorphous mass composed of their corpses at the bottom of my collecting glasses. It is, in fact, essential for the preservation of many of the gelatinous surface organisms that they should, in this country, in the absence of an apparatus by which they can be supplied with a constant stream of cool water, be at once treated with the necessary fixing and preservative re-agents ; but the management of the requisite processes is by no means an easy matter in the limited space afforded by a native canoe. The suggestion made by Haeckel that the death and decomposition of the delicate organisms might be prevented by placing them in vessels cooled by ice is, without doubt, an excellent one ; but unfortunately, ice cannot, as a rule, be procured in out-of-the-way places where one most requires it.

Among the pelagic organisms which I have collected over the coral reefs in the Gulf of Manaar may be mentioned various small *Medusæ*, *Beroe*, *Cydippe*, *Bolina* (present one morning in such abundance that the net became instantly filled with a thick jelly), dense crowds of *Copepod* and *Schizopod* crustaceans sometimes rendering the surface of the water milky ; *Zoæa*, *Phyllosoma*, and *Alima* larvæ ; violet-blue *Janthinæ;* and *Styliola acicula*, a pteropod Mollusc, whose dead glassy shells are very abundant in deposits from the sea bottom. Less frequently met with were young *Cephalopods*, of which the adults, as well as an *Annelid* (*Nereis*) obtained by digging deep holes in the sand, are extensively used as bait by the fishermen; *Salpæ;* and the ova and young of fishes. Floating, too, on the surface of the water, and conspicuous by their bright colouring, were various *Siphonophora—Physalia* (the Portuguese Man-of-war), *Velella* with its sun-dial-like crest, and *Porpita* with its exquisitely marked disc. Many minute pelagic animals were obtained by shaking in a tumbler of water the marine *Algæ* which were floating over or living on the reefs, and of which the most conspicuous were *Sargassum vulgare* (Fig 1) and *Padina pavonia* (peacock's tail). These pelagic organisms, from which the main food-supply of the coral polyps is probably derived, were far more abundant and varied over the Pámban reef during my visit to Rámésvaram Island in 1886 than in 1888 : and this is probably to be explained by the fact that, in the former year, there was but little wind, and the water was so clear that, in the early morning before the gentle day breeze set in, the individual corals could be clearly distinguished as I rowed over the reef;

FIG. 1. *Sargassum Vulgare.*

whereas in the latter year there was generally a strong southwest wind blowing, and a rapid current running through the Pámban Pass carrying with it sediment in suspension, which rendered the water turbid: and, as is known, a pure and transparent condition of the water is the first and indispensable condition for the life of many marine creatures, especially those of the coast. Moreover, the ripple on the surface probably drove the pelagic animals into deeper water, which I did not explore in search of them. On calm mornings, when the surface has been teeming with small *Medusæ*, I have seen the living organisms and their dead gelatinous remains adhering in large quantities to the surface of living corals with their tentacles expanded, which were brought up for me by my divers. There has been a noticeable absence of big jellyfishes during my visits to Rámésvaram Island. Only, in fact, during the last few days of my stay on the island in 1889

MAINLAND AND RÁMÉSVARAM ISLAND

did I see a few large *Rhizostomas* (called by the natives *sori*= nettles), floating over the reef or washed on shore. Phosphorescence, too, I have never seen well marked in the Gulf of Manaar, the sight of an occasional luminous flash from a pelagic crustacean being the poor reward of night vigils.

The island of Rámésvaram, which is visited during the course of the year by enormous numbers of Hindu pilgrims to the celebrated temple, is separated from the mainland by the Pámban Pass, which connects Palk's Strait with the north end of the Gulf of Manaar, and is 1,350 yards in width.[1] The depths in the channel range from $10\frac{1}{4}$ to 15 feet at low water, but it shoals up very suddenly on both sides, so that great care is necessary in navigating vessels through. On the west side of the pass is the great dam, consisting of large masses of sandstone, all having a more or less flat surface, and which formerly were part of a causeway extending from Rámésvaram Island across to the mainland. The remains of this causeway are still visible on the main road from Pámban to the town of Rámésvaram.

According to the folk-lore of the Hindus, the so-called bridge, which formerly connected Rámésvaram Island with Ceylon, was built by an army of monkeys when Ráma made war against Rávana, who had carried off his wife Sita to the island of Lanka (Ceylon), and as Mr. Bruce Foote observes:[2]

"The series of large flat blocks of sandstone so strongly resemble a series of gigantic stepping-stones, that it is impossible to wonder at the imagination of the author or (in analogy with the Homeric epos) authors of the Ramayana that the rocky ridge was really an old causeway of human construction."

In connection with the building of the reef a story goes to the effect that the common South Indian squirrel (*Sciurus palmarum*) used to help the monkeys by rolling in the sand on the shore, so as to collect it in its thick hairy coat, and then depositing it between the piled up stones, so as to cement them together. At which service Ráma was so

---

[1] The following account of the Pámban Pass is mainly based on Extracts from Hydrographic notice, derived from survey and remarks furnished by Mr. Morris Chapman, Assistant Superintendent, Marine Survey of India, 1878.

[2] *Mem. Geol. Surv., Ind.*, vol. XX, 1883.

pleased that he stroked the squirrel on the back, which has, ever since, borne the finger marks.

Writing in 1821 concerning Adam's Bridge, Davy observes [1] that:

"No one who looks at a map and notices the little distance (about 17 miles) between the nearest point of the island (Ceylon) and continent, and how, by the chain of rocks and sand-banks commonly called Adam's Bridge, they are still imperfectly connected, can entertain much doubt that the connection was once perfect. This inquiry is more curious than useful. It would be much more useful to endeavour to complete that which nature has begun, and to make the channel, which is now obstructed and dangerous, clear and safe, and fit for the purposes of coast navigation. If, on examination, sandstone and coral rock should be found constituting part of Adam's Bridge instead of primitive rock, one necessary inference is that the channel, at whatever period formed, was once deeper and more open than it is at present, and another inference is that, in process of time, it will be closed up, and Ceylon joined to the continent."

Tradition runs to the effect that, at the time of the disruption of Rámésvaram Island from the mainland on the one side and Ceylon on the other, the cows became prisoners on the island, and being unable, like the cows at Cape Cod, which are fed on herring's heads, to adapt themselves to a fish diet, took to living on sea weeds, and have become, by degrees, converted into diminutive "metamorphosed cows," which may still be seen grazing on the shore. This story is based on the fact that portions of the skulls of cats and dogs, including the articulated temporal, parietal, and occipital bones, which are sometimes picked up on the beach, bear a rude resemblance to the skull of a cow, the horns being represented by the zygoma.

When I was staying at Pámban in 1888, a bucket dredger was at work in the Pass, and from the mud brought up by it, I obtained many small Crustacea, Echinoderms (chiefly *Laganum depressum* and *Fibularia volra*), Mollusca (of which *Leda mauritiana* was one of the most abundant) including great quantities of the little *Avicula vexillum*, which was formerly mistaken for the young of the pearl oyster, a single Gephyrean (*Sipunculus* sp.), *Branchiostoma*

---

[1] *Travels in Ceylon.*

Amphioxus), and many fragments of a small *Fungia*, which must be very plentiful, but of which no perfect specimen was obtained.

Southward of the Pámban Pass are three islands, Pulli, Pullivausol, and Coorisuddy, completely encircled by an irregular coral reef, the whole forming a natural breakwater protecting the pass and the channels leading to it from the violence of the south-west winds. The space between the northern edge of this reef and the pass forms a fine sheltered anchorage for vessels of light draft in all weathers. The deepest water between the above islands and the Pass is immediately south of Coorisuddy, and is called the basin, over which there is an average depth of 18 feet, but in one spot there is a depth of 21 feet. This basin is, however, very narrow, being simply a hole scoured out by the action of the water in rushing through the pass: and, consequently, is of little value to ships, as it has the pass to the northward of it with only 10 feet, and the sand-bank channel to the southward with only 9½ feet at low water.

The tides are very irregular at Pámban, the rise and fall being much affected by the winds. The average springs rise 3 feet; but, during neaps, sometimes for 48 hours, there is frequently only a rise and fall of 1 or 2 inches. The currents are generally influenced by, and strong in proportion to the force of the wind. Through the Pámban Pass the current frequently attains to a velocity of from 5 to 6 knots an hour, rendering it at times difficult even to take full-powered steamers through. During the north-east monsoon the current sets to the north through the pass. The only months in which a real tidal current is noticeable are March, April, and October, when it generally sets six hours each way.

No records of the temperature of the water over the reef are extant, and, as my visits have always been at the same season, extending over only a few weeks of the year, the temperature observations which I have made are practically of no value. The following table, however, shows the maximum and minimum and monthly range, recorded at the Pámban Marine office in the shade at 10 A.M. and 4 P.M. during the twelve months from April 1st, 1888, to March 31st, 1889. The range of temperature during that period will be seen to be from 76° to 92°, *i.e.*, 16°:—

	Min.	Max.	Range.
April 1888	81°	92°	11°
May ,,	79°	91°	12°
June ,,	84°	88°	4°
July ,,	84°	89°	5°
August ,,	84°	88°	4°
September ,,	84°	89°	5°
October ,,	78°	89°	11°
November ,,	78°	89°	11°
December ,,	77°	86°	9°
January 1889	76°	81°	5°
February ,,	80°	88°	8°
March ,,	82°	92°	10°

The town of Pámban is situated on the western extremity of the island, and lies to the west and south-west of the lighthouse, built on the top of a sand-hill, at the foot of which is a good example of sand-rock, *i.e.*, a mass of fine sand, which has become compacted by the action of wind and spray, so as to form a stratified friable rock exposed amid the surrounding loose blown sand. With the exception of the Port officer's house and a few others, the houses consist principally of huts made of *cajan* leaves. The native population is mainly made up of boatmen and fishermen, some of whom find employment in carrying coolies over to Ceylon, and others in ferrying the pilgrims from the mainland to the island. There are also a large number of coolies, who are engaged in hauling vessels through the Pass when the wind is adverse.

Pámban boasts of a ruined fort built by the Dutch during the Dutch occupation of the island, over which I was taken by a native guide, who pointed out as objects of interest some stone cannon-balls, battered dredge-buckets of modern construction, and some barrels of fuse lying mouldering from age in what he termed a *conjee* (rice) house, a damp, ill-ventilated building, wherein, at some period at which the Public Works Department was engaged on works in the island, the recalcitrant sapper used to be placed in confinement on a sedative *conjee* and water diet.

As regards the food-supply at Pámban, beef and mutton are not easily procurable, goat, long-legged and emaciated, being the principal animal supplied. Fowls and native vegetables can always be obtained in the bazár. The local oggs possess a peculiar flavour which is attributed to the fact that the fowls feed partly on fish, affording an example of *polyphagy*, which reminds one of the observation of John Hunter, that a species of Gull (*Larus tridactylus*) which, though commonly feeding on fish, and having its stomach adapted to flesh diet, can also live on grain. Another species of Gull (*Larus argentatus*) is said to live in the Shetland Islands on grain in the summer and on fish in the winter.

The fish supply at Pámban is very plentiful, and a visit to the ill-smelling fish bazár always showed an abundance of fish, unappetising Cephalopods, and Crustacea (*Neptunus pelagicus, Scylla serrata, &c.*) for sale. During my visit in 1889 the following food-fishes were obtained either by means of a drag-net or from the bazár [1]:—

ELASMOBRANCHII (Sharks and rays).

Carcharias. sp. juv.	Trygon uarnak.
Zygæua malleus.	Myliobatis nieuhofii.

TELEOSTEI (Bony fishes).

Lates calcarifer.	Teuthis oramin.
Lutianus rivulatus.	Caranx speciosus.
Lutianus roseus.	Caranx ire.
Therapon theraps.	Equula edentula.
Pristipoma hasta.	Sillago sihama.
Scolopsis. sp.	Mugil speigleri.
Gerres oyena.	Cynoglossus macrolepidotus.
Drepane punctata.	Arius thalassinus.
Scatophagus argus.	Saurida tumbil.
Upeneoides tragula.	Hemiramphus xanthopterus.
Upeneus indicus.	Clupea. sp.
Lethrinus nebulosus.	Pellona leschenhaulti.

My head-quarters on the island have been mainly fixed at the zemindar's bungalow, situated on the top of a sand-hill near the Pámban light-house, which, were it more easy of access from Madras, would make an excellent marine biological station. But I have occasionally pitched my camp at a muntapum on the shore at Rámésvaram, close to the spot where the pilgrims go through certain mysterious ceremonies and ablutions.

---

[1] The technical names given are those used in Day's *Fishes of India*.

The verandah of the zemindar's bungalow affords a good spot for the study of the common animals and birds of the island. The former consist mainly of ill-conditioned pariah dogs; goats trying to extract the requisite amount of foodstuffs for the maintenance of life from dried palmyra leaves and spiny *Acacia planifrons*, the spines of which serve as no protection against the attacks of these hard-mouthed herbivorous mammals; and donkeys suffering from motor paresis of their hind limbs. The shrill voiced palm squirrel and musk-rat infested the bungalow, and a friendly mungoos made repeated visits when I was at breakfast. Of birds, the splendid but shameless crow[1] made continual raids on my specimens drying in the sun; and parakeets screaming in a neighbouring Bo tree, and screech-owls making night hideous with their domestic quarrels proved a constant source of irritation.

During my stay on the island in 1886 the following birds were shot by my shikaree [2]:—

17. Tinnunculus alaudarius. Kestrel.
23. Micronisus badius. Shikra.
76. Athene brama. Spotted owlet.
117. Merops viridis. Indian bee-eater.
149. Palæornis rosa. Rose-headed parakeet.
180. Brachypternus aurantius. Golden backed woodpecker.
197. Xantholæma indica. Crimson-breasted barbet.
205. Hierococcyx varius. Common hawk cuckoo.
212. Coccystes melanoleucos. Piedcrested cuckoo.
217. Centropus rufipennis. Common coucal.
255. Upupa nigripennis. Indian hoopoe.
257. Lanius erythronotus. Rufous-backed shrike.
260. Lanius hardwickii. Bay-backed shrike.
278. Dicrurus macrocercus. Common drongo shrike.
433. Malacocircus griseus. White-headed babbler.
452. Ixos luteolus. White-browed bush bulbul.
462. Pycnonotus hæmorhous. Madras bulbul.
467. Iora zeylonica. Black-headed green bulbul.
475. Copsychus saularis. Magpie robin.
660. Corvus culminatus. Indian corby.
663. Corvus splendens. Indian crow.
684. Acridotheres tristis. Common myna.

---

[1] *Corvus splendens*, vel. *impudicus*.
[2] The technical names given correspond, as do the numbers, to those used in Jerdon's *Birds of India*.

687. Temenuchus pagodarum. Black-headed myna.
795. Turtur suratensis. Spotted dove.
884. Tringa minuta. Little stint.
944. Phænicopterus roseus. Flamingo.
980. Xema brunnicephala. Brown-headed gull.
985. Sterna seena. Large river tern.

On the sandy shore of Shingle Island, one of the islands which intervenes between Rámésvaram Island and the mainland, which is overgrown with long grass reaching in some places to a height of 6 feet, my friend Mr. J. R. Henderson saw, in early June, hundreds of a doubtful species of Tern (?) and a few of the large river Tern (*Sterna seena*). Of these the latter laid a single egg in a tunnel excavated among the matted roots of the grass, and artfully concealed from view. The former laid a single egg in a hole scooped out in the sand near the water's edge, where the grass was either very short or absent; and the eggs were easily missed owing to the resemblance between their colour and that of the sand, which affords an example of the adaptation of the colouring of eggs to their natural surroundings for the purpose of concealment, according to the principle of protective colouration. In July 1888 the shores of Coorisuddy Island [1] were in possession of an army of occupation of flamingoes, which were, no doubt, feeding on worms and burrowing crabs.

The insect would, apart from the irrepressible ants and mosquitoes, is only poorly represented on Rámésvaram Island, and of *Lepidoptera* the most conspicuous was *Papilio* (*Menelaides*) *hector*, flying swift-winged along the shore or far out at sea. The following *Lepidoptera* were captured in July 1889 :—

Mycalesis mineus, *Linn.*
Melanitis leda, *Linn.*
Tarucus plinius, *Fabr.*
Catochrysops strabo, *Fabr.*
Catopsilia crocale, *Cramer.*

Catopsilia catilla, *Cramer.*
Terias hecabe, *Linn.*
Papilio hector, *Linn.*
Papilio erithonius, *Cramer.*

---

[1] The following botanical specimens were collected by me on Coorisuddy Island :—*Dodonæa viscosa, Eugenia jambolana, Pomphis acidula, Oldenlandia umbellata* (Indian madder or chay root), *Vernonia cinerea, Launæa pinnatifida, Salvadora persica, Enicostema littorale, Ipomæa biloba* (Goat's foot), *Clerodendron inerme, Boerhaavia diffusa* (spreading hog weed), *Aerua javanica, Suæda monoica, Euphorbia corrigioloides, Euphorbia thymifolia, Phyllanthus polyphyllus, Pandanus odoratissimus* (screw-pine), and *Cynodon dactylon* (huriallee grass).

Though I came across none myself, I was shown a bottle containing a collection of scorpions, which had been caught at Pámban; and the big spider (*Mygale*), which is accused of causing the death of sheep and goats by poisoning them on the muzzle, has been captured by Mr. Henderson in the zemindar's bungalow.

Commencing near the zemindar's bungalow and extending for some distance along the north coast of the island, is a fossil coral reef, which I cannot do better than describe in Mr. Bruce Foote's words [1]:

"The upraised reef," he says, "is a striking feature of the north coast of Rámésvaram Island, and is worthy of much closer study than the time at my disposal enabled me to bestow upon it. It shows best along the beach beginning a couple of hundred yards west of the zemindar's bungalow, where it forms a little irregular scarp about a yard or 4 feet high, against the roof of which the waves break in rough weather. Of its true coral reef origin there can be no doubt, as in many places the main mass of the rock consists of great globular meandroid corals, or of huge cups of a species of porites which, beyond being bleached by weather action, are very slightly altered, and still remain in the position in which they originally grew. The base of the reef is not exposed, as far as I could ascertain, not having been sufficiently upraised along the beach; but in a well-section a little to the south of the Gandhamána Parvattam Chattram the thickness of the coral reef exposed above the surface of the water is at least 10 feet, and probably much more. The great swampy flat forming the northern lobe, as it were, of Rámésvaram Island, consists, I believe, entirely of this upraised reef hidden only by a thin coating of alluvium, or the water of the brackish lagoons which cover the major part of the surface, but do not form a continuous sheet of water as shown in the map.

"I came across masses of coral protruding at intervals through the alluvium in the very centre of the flats north-westward of the great sand-hill crowned by the chattram just named. The raised reef is very well seen to the north-eastward of Rámésvaram town, where it forms a miniature cliff from 3 to 4 or possibly 5 feet high, and continuing along the coast after the latter turns and trends to north-west. Time did not admit of my actually following it up to Pesausee Moondel Point, but I went to within a mile of the point, and could see no change of character of the coast line on examination through a strong

---

[1] *Mem. Geol. Surv., Ind.*, vol. XX, 1883.

Fossil Reef at Pámban.

field-glass. The raised reef shows strongly also along the western side of the flat northwards of Ariangundu. The south side of the reef is, along the north coast, completely covered up by the great spreads of blown sands which occupy the greater part of the surface of the island. On the east side of the island the reef does not extend close up to the great temple, but stops short abruptly about 300 yards to the north-east, and does not re-appear on the coast of the bay south of the temple. South of Pámban town also there were no signs of any upraised coral, nor could I see any indication eastward along the south coast, as far as the eye could reach from Coondacaul Moondel Point, while the great south-east spit terminating at the point called Thunnuscody is covered by a double ridge of great blown sand-hills. An important series of trial sinkings made by the Port officer at Pámban right across the island, from north to south, about 2 miles east of the town, in order to test the feasibility of the proposed ship canal, did not reveal any southerly extension of the raised reef. The probability is that it forms a mere narrow strip along the beach from Pámban to Ariangundu, but widens out thence to the north-eastward to form the northern lobe of the island.

"Parts of the reef lying between collections (colonies as it were) of the great globular or cup-shaped coral masses form a coarse sandstone made up of broken coral, shells, and sand (mostly silicious) a typical coral sandstone.

"At the Pámban end of the raised reef it shows a slight northerly dip, and masses of dead coral, apparently *in situ*, protrude through the sand below high-water mark. Reefs of living coral fringe the present coast, but these I was unable to examine, so cannot say whether the corals now growing there are specifically allied to those which formed the reef now upraised, but all the mollusca and crustacea I found occurring fossil in the latter belong to species now living in the surrounding sea."

Mr. Bruce Foote writes further :—

"It is quite evident from the occurrence of the old coral reef on Rámésvaram Island that the latter must have been upraised several feet within a comparatively recent period, but unfortunately there are no data by which to calculate the exact amount of the upheaval. The upheaval which affected Rámésvaram Island doubtless affected the adjoining mainland, and, by upraising the coast, exposed the sandstones, which have been described above as forming a low wall-like cliff bordering the beach as if by a built quay."

A good example of a sandstone quay wall is to be seen on the mainland between the great dam and Muntapum.[1]

In a letter to me Dr. Johannes Walther, of the University of Jena, who made a short visit to Rámésvaram Island early in 1889, writes concerning the fossil reef:—

"Das fossile riff beginnt direct unter dem bungalow, und lässt sich um die ganze nordküste bis nach Ramesvaram verfolgen. Es ist wundervoll. Porites 4 m. dick, und viele andere formen. Bei Ramesvaram ist ein sehr schönes fossiles Lithothamnium lager, und interessante metamorphosen des subfossilen riffes."

Possessing only very scanty geological knowledge, I am unable to deal satisfactorily with the fossil reef, which will, I trust, receive full justice from Dr. Walther.[2] Commencing, as already stated, near the zemiudar's bungalow it forms a wall exposed to a height of 3 or 4 feet above the sandy shore in which it is imbedded, and extending, almost without interruption, for a distance of a quarter of a mile, after which it becomes covered over with loose sand, and is exposed only at intervals. The main mass of this wall, as also of the big detached coral blocks which intervene between it and the sea, and are washed by high tides, is built up of enormous blocks of *Porites*, one of which, isolated from neighbouring blocks, has a diameter of 12 feet. That these blocks are imbedded as they grew is shown not only by their reef-like appearance, but also by their upright position, the vertical columns of many of the blocks bearing testimony to the fact that they have not been cast up by the waves at random, like the big coral fragments which are exposed at low tide, and lie irregularly in all possible unnatural positions. The calices on the surface of the fossil corals are

---

[1] North of Kílakarai, a town on the coast southwest of Rámésvaram Island, a very perfect wall of sandstone extends for some distance along the shore, in the loose sand covering which many copper coins—Roman, Chola, Pándyan, Dutch, Indo-French, &c.—have been recently found; and a theory has been started that this spot is the site of an old Pándyan city. The area which intervenes between the fringing coral reef and the sloping shore at Kílakarai, and is uncovered by water at low tide, is covered by an extensive green carpet formed by a dense growth of *Zoanthi* agglutinated together by damp sand, among which small isolated *Madrepores* live, though periodically exposed to the heat of the sun. Opposite the town of Kílakarai there is a wide gap in the reef, through which small sailing boats can pass into the shallow harbour within the reef, on which the force of the waves is broken.

[2] *Vide* a Report by Dr. Walther in the *Verhandlungen der Gesellschaft für Erdkunde zu Berlin*, 1889, No. 7.

either perfectly distinct over large areas, so as to render their identity certain, or, especially in the case of the blocks which are still exposed to wave action, worn away, or concealed by a crystalline incrustation. Imbedded in cavities in the *Porites*, once bored and occupied by the living mollusc animal, are immense numbers of the shells of the lithodomous *Venerupis macrophylla*, which abounds on the living reef at the present day. The *Porites* are frequently capped by *Astræans*, which are also found firmly fixed to their lateral aspect. Less commonly they are incrusted with *Mæandrinas* (*Cæloria*), which, like the *Astræans*, also form solid isolated blocks, but of far smaller size than the *Porites*. The blocks are, for the most part, covered on their upper surface by a crust of thick compact laminated sand-rock, imbedded within which are the shells of mollusca—*Cardium*, *Arca*, *Turbo*, *Cerithium*, *Spondylus*, &c. I have also found several carapaces of fossil crustacea, whose species I am unable to identify. At the commencement of the reef, *i.e.*, at the end nearest to the bungalow, the sand-rock is arranged in a succession of layers with a dip seawards, and forms an incrusting layer about 8 inches thick. A little further on the reef has a terraced appearance, an upper terrace being formed by sand-rock horizontally stratified, exposed to a height of 18 inches, and supported by underlying *Porites*, *Astræa*, *Cæloria*, and *Turbinaria*; and a lower terrace formed by a flat-topped mass of *Porites*, about 9 yards in length, covered with loose sand. Not the least interesting feature of the coral wall is the presence of a bank of *Madrepores*, extending over a length of 8 yards at a higher level than the *Porites*, and evidently still placed as they originally grew, their radiating branches spreading outwards from the base, and forming a broad flat surface, which affords support to a thick superjacent layer of consolidated sand-rock. The maximum height of the Madrepores above the loose shore sand is 18 inches, and they clearly form a portion of a bank, such as may be seen spreading over considerable areas on the living reef on a calm day.

As one looks out to sea from the Pámban beach at low water on a breezy day, three distinct zones can be clearly distinguished, viz. :—(1) commencing about three-quarters of a mile from the shore, and extending to the horizon, clear blue water separated by a sharp line of demarcation from (2) a zone discoloured by sediment in suspension carried by the current through the Pámban Pass. This zone, in which the living corals flourish in spite of the current, sometimes running at the rate of 7 to 8 knots per hour, to which they are exposed, terminates at the sharply defined land face of the reef,[1] the corals of which, constantly bathed by water and never exposed above the surface, act as a natural breakwater which breaks the force of the waves, so that, at high tide, the shallow water between the reef and the shore is smooth. The land face of the reef is made up almost entirely of *Madrepores*, amid a perfect forest of arborescent sea weeds and fleshy *Alcyonians* which, as one rows over the reef on a bright still morning, can be easily recognised as large snow-white patches. Other genera—*Porites, Cœloria, Turbinaria*, &c., occur in deeper water. (3) There is a zone, about 40 yards in breadth, between the reef and the shore, which is covered by water at high tide, but completely exposed at low tide, and made up of dead coral blocks, fragments, and débris, among which branches of worn *Madrepores* are most conspicuous, broken off or rolled along from the reef, and covered with low-growing clumps of brown and green sea weeds, and enclosing shallow pools in which "coral fishes" of brilliant hue may be seen, and colonies of *Cerithia* leaving in their wake a characteristic track. Many of the larger coral blocks are extensively worn by the process of solution, or eroded by boring mollusca and other animals. Among the crevices of the eroded corals various crustacea (*Gonodactylus, Pilumnus*, &c.), find a home; and crawling on their surface, which is frequently covered by erect or sessile encrusting sponges, or hidden beneath them, Annelids (*Amphinome, Nereis*, &c.), and bright-colored Planarians may be found.

From the Pámban beach the sea bottom slopes very gradually to a depth of 20 to 26 feet at a distance of three-quarters of a mile from the shore. Between the Kathoo Vallimooni Reef, marked on the survey chart as being "partially dry at low water spring tides," and the spit of mainland which terminates at Point Rámen a boat passage has been carved out by natural processes. North of Rámesvaram Island the living coral reef formation is stated by the

---

[1] In the recently issued third edition of Darwin's *Structure and Distribution of Coral Reefs*, the reefs of the Madras Coast, of the Gulf of Mannar and the northern part of Ceylon are not indicated on the map showing the distribution of coral reefs because as Professor Bonny says (p. 247):— "The sea off the northern part of Ceylon is exceedingly shallow, and, therefore, I have not coloured the reefs which partially fringe portions of the shores and the adjoining islets, as well as the Indian promontory of Madura."

local fishermen, in answer to independent inquiries by Mr. Bruce Foote and myself, to extend only as far as Pillay Mudum, 7 miles south-east of the Vigai river, which, though easily crossed on foot in the dry season, is in high flood during the monsoon, and, for about a fortnight in the year, impassable even on a raft.

Piled up over a limited area at the base of the fossil reef were masses and fragments of pumice [1] encrusted with *Polyzoa, Chamæ,* tubes of tubicolous worms, *Balani,* young pearl oysters, &c., and dislodged in the first instance, in all probability, from the volcano of Krakatoa during the great eruption of 1883, a curious result of which has been that, in the district of Charingin, which was depopulated by the tidal wave during the outburst, tigers have increased so enormously in number that the Government reward for killing them has been fixed at 200 guilders each.

Washed on shore by the waves, protecting the upper surface of the dead corals, or brought up for me from the sea bottom by my divers, were nodular calcareous *Algæ,* which, from microscopical examination, I find to be identical with those which were dredged off the town of Negombo in Ceylon by Captain Cawne Warren, and reported on by Mr. H. J. Carter.[2] "The specimens," says that authority, "consist of calcareous nodules of different sizes, which may be said to originate, in the first instance, in the agglutination of a little sea bottom by some organism into a transportable mass which, increasing after the same manner as it is currented about, may finally attain almost unlimited dimensions. They are, therefore, compounded of all sorts of invertebrate animals, whose embryoes, swimming about in every direction, find them, although still free and detached, of sufficient weight and solidity to offer a convenient position for development, and hence the number of species in and about them . . . . Perhaps no family of organisms has entered into their composition or increased their solidity more than calcareous Algæ (*Melobesiæ*) which, in successively laminated or nulliporoid growths, have rendered these

---

[1] "The fragments of pumice thrown up into the ocean during far distant sub-marine eruptions, or washed down from volcanic lands, are at all times to be found floating about the surface of the sea, and there being cast upon the newly formed islet produce by their disintegration the clayey materials for the formation of a soil, the red earth of coral islands." Murray, Royal Institution, March 16, 1888.

[2] *Ann. Mag. Nat. Hist.,* June 1880.

nodules almost solid throughout, or covered with short, thick, nulliporiform processes . . . . . . Next to the part which the *Melobesiæ* have taken in their formation may be mentioned the sessile *Foraminifera*, and these have, in turn, been overgrown, in many instances, by *Polyzoa*."

Specimens have been picked up on shore both by Mr. Bruce Foote and myself of a curious body, the nature of which has given rise to some discussion, and is still *sub judice*. One of them was exhibited at the Linnean Society, and Dr. Anderson and Mr. Dendy were inclined to regard it as, possibly, the consolidated roe of a fish ; whereas Professor Charles Stewart was of opinion that it was a vegetable structure, his opinion being based on the examination of microscopical preparations which he demonstrated to me when I was recently in Europe.

Among other specimens collected on the Pámban beach I may mention the complex tubular skeletons of the Chætopod *Filograna*, and large blocks of drift wood bored by the mollusca *Teredo* and *Parapholas*, of which the latter has recently destroyed the bottom of the local port gig.

The Indian fin-whale (*Balænoptera indica*) has been known to accompany vessels in the Gulf of Manaar, and I have seen one close to a steamer in which I was rounding Cape Comorin. It is related that, some years ago, the schooner *Abdul Rámán*, which was at anchor close to Pámban, was suddenly released from her moorings, and towed out to sea to a distance of several miles by some invisible agent. A few days afterwards the carcase of a whale was cast on shore, and the theory was that this whale was the cause of the involuntary cruise, it having been tempted out of curiosity to examine the ship, in whose grapnel it is supposed to have been caught, and to have taken the steamer in tow until it liberated itself.

The phytophagous Sirenian *Halicore dugong* (the dugong), which is said [1] to be found in the salt water inlets of South Malabar, feeding on the vegetable matter about the rocks and basking and sleeping in the morning sun, is according to Emerson Tennent [2] attracted in numbers to the inlet from the Bay of Calpentyn on the west coast of Ceylon to Adam's Bridge by the still water and the abundance of marine algæ in this part of the Gulf of Manaar. It is of an extremely

---

[1] Jerdon, *Mammals of India*.   [2] *Ceylon*, vol. II, 1860.

shy disposition, and I have never seen it myself, though I have heard of dead carcases being thrown up on the Pámban beach, or living specimens being caught in the fishing nets. One was, in fact, caught, together with a young one, the day before my arrival at Pámban in 1889, and promptly sold for food, as it is considered a great delicacy. There is a tradition among the natives that a box of money was found in the stomach of a dugong which was cut up in the Pámban bazár some years ago; and an official is now always invited to be present at the examination of the stomach contents, so that the possessors of the carcase may not be punished under the Treasure Trove Act for concealing treasure. But the stomach contents invariably prove to be green sea-weed. The fat of the dugong is believed to be efficacious in the treatment of dysentery, and is administered in the form of sweetmeats, or used instead of *ghee* in the preparation of food. The skeleton of a female dugong in the Madras Museum shows, encased in the upper jaw, the functionless teeth, the blunt points of which are, during life, covered by a fleshy lip forming a snout. The female is described by Tennent (*op. cit.*) when suckling her young, as holding it to her breast with one flipper, while swimming with the other, holding the heads of both above water, and, when disturbed, suddenly diving and displaying her fish-like tail.

The edible turtle (*Chelone mydas*) which I have seen carrying the cirrhiped *Chelonobia testudinaria*[1] and the pearl oyster attached by its byssus to the carapace, is very abundant in the shallow water near the sandy shores of the islands in the vicinity of Rámésvaram, on which the female lays her eggs. A large specimen, whose skeleton has been preserved, was purchased for eight annas on the understanding that the vendor should have the flesh as a perquisite. The process of removal of the edible portions fat, flesh, and viscera was not a pleasant operation to witness. The victim was placed on its back, and secured by ropes which did not prevent demonstrative flapping of its fins during the operation, skilfully performed with a carving knife, of removal of the breast-plate so as to display the internal organs, which were removed together with their investing fat; the pulsations of the heart, which was removed last of all, the snapping of

---

[1] I have also seen parasitic pedunculated cirrhipeds attached to the skin of a sea-snake (*Hydrophis*), the gills of *Neptunus pelagicus*, and the antennæ of *Palinurus dasypus*.

the jaws, the plaintive expression of the eyes, and general indications of discomfort forming a ghastly spectacle not easily to be forgotten. The flesh of the edible turtle is described by Tennent as being sold piecemeal in the market place at Jaffna while the animal is still alive, each customer being served with any part selected which is cut off and sold by weight; and Darwin, referring to the gigantic tortoise of the Galapagos Archipelago, says that, when a tortoise is caught, a slit is made in the skin near the tail, so as to see whether the fat under the dorsal plate is thick. If it is not, the animal is liberated, and it is said to soon recover from the minor surgical operation.

A single specimen of the land tortoise (*Nicoria trijuga*), found at the foot of a tree on the sandy soil outside the town of Pámban, was brought to me for sale. The land snakes of the island are represented, so far as I know, by *Lycodon aulicus* and *Tropidonotus stolatus*, of which the latter bit Mr. Henderson's native servant in the foot, causing great torture until he was assured that it was not a *toxicophidian*. Batrachians I have not seen on the island, but the existence of *Rana hexadactyla*, which is, I am told, eaten in the Indo-French possessions, was made evident by its nocturnal croaking in a tank near the bungalow.

One of the edible *Holothurians*[1] (*Trepangs* or *Béches-de-mer*) is very abundant in the mud on the south shore at Pámban, and in the vicinity of Rámésvaram, at both which places it is prepared for exportation to Penang and Singapore. The process of preparation, which is not an appetising one to watch, is as follows:—

The *Holothurians* are collected as they lie in the mud at low water, and placed in a chaldron which is heated by a charcoal fire. As the temperature rises in the chaldron the still living animals commit suicide by the convenient process of ejecting their digestive apparatus, &c., and become reduced to empty leathery sacs which, by loss of water consequent on the temperature to which they are exposed, shrivel considerably. At the end of twenty minutes or half an hour the boiling process is stopped, and the shrivelled animals are buried in the sand until the following morning, when the boiling process is repeated. Finally, they are arranged according to their size, and are ready for shipment.

---

[1] *Holothuria marmorata.*

As regards the question [1] whether *Holothurians* live on living coral or obtain nutriment from swallowing the sand and detrital material, the two most abundant species in the Gulf of Manaar (*H. atra* and *H. marmorata*) live, not on the reef, but on the muddy bottom between the reef and the shore, which is frequently uncovered at low tide. From repeated examination of the contents of their alimentary canal, I have been unable to find any evidence that they have been feeding on living coral, the swallowed materials consisting, for the most part, of sand, coral débris, small Mollusca, Alcyonian spicules, and sea weeds.

---

[1] *Vide* Darwin, *Coral Reefs*, 3rd edition, 1889, p. 20.

VI.—FAUNA OF THE GULF OF MANAAR.

VI.—FAUNA OF THE GULF OF MANAAR.

THIS report upon the Fauna of the Indian side of the Gulf of Manaar is composed of lists of those species which have been obtained by myself from Rámésvaram and the neighbouring islands, and Tuticorin, including the pearl banks.

I take this opportunity of expressing my hearty thanks to Mr. Arthur Dendy, Professor Jeffrey Bell, Mr. J. R. Henderson, and Mr. Edgar A. Smith, for the great assistance which they have rendered in connection with the collections of sponges, echinoderms, crustacea, and mollusca.

The report must be regarded as a preliminary one, and I have made no attempt to classify the worms, nudibranchs, tunicata, etc., which await identification.

PORIFERA.

The sponges recorded below were collected by me either in the neighbourhood of Rámésvaram Island or at Tuticorin, and sent to Mr. Arthur Dendy at the British Museum, Natural History, by whom they are described in detail in the Annals and Magazine of Natural History, September 1887, and February 1889.

As regards the first collection, which was made at Rámésvaram, Mr. Dendy writes:—

"The collection is of exceptional interest, owing to the fact that it is the first which has been obtained from this particular locality. Indeed our knowledge of the sponge-fauna of the entire Indian Ocean is extremely deficient. This deficiency is almost certainly due to want of investigation rather than to any actual scarcity of sponges. Mr. Ridley and I have already pointed out, in our report on the Monaxonida collected by H. M. S. Challenger, that 'this little-known field will probably yield a rich harvest to whoever has the good luck to thoroughly investigate it;' and this statement is amply borne out by Mr. Thurston's researches.

"The best known locality for sponges in the Indian Ocean is undoubtedly Ceylon; Bowerbank, Gray, and Carter have all written upon the sponge-fauna of this particular district, and the sponge-fauna of Madras, in so far as is evidenced by the material at my disposal, bears a striking resemblance to it. Thus, out of the ten determinable species from Madras, four, viz., *Halichondria panicea* (a cosmopolitan species), *Axinella donnani*, *Hircinia clathrata*, and *Hircinia vallata*, have already been recorded from the neighbourhood of Ceylon.

"There can be no doubt that the present collection was obtained in shallow or moderately shallow water, although there is no record of the depth. Species with a strong development of spongin in the skeleton-fibre predominate, as might have been safely predicted from the climatic conditions of the locality."

The majority of the sponges, as will be seen, belong to the Monaxonida, which "comprise by far the most commonly met with and abundant of all sponges. They occur in greater or less profusion in all parts of the world, but are more especially shallow-water forms. They may be collected between tide-marks almost anywhere."[1]

None of the Gulf of Manaar sponges, which I have collected from between tide-marks up to 11 fathoms, are of any commercial value.[2] The colours of many of them are very bright, but soon fade or change when the sponge is dried or immersed in alcohol.

The following list includes only a portion of my collection, many of the sponges still awaiting identification. Of the 31 species recorded there are 18 new species, and 2 new varieties of previously recorded species, viz., *Pachychalina multiformis* and *Ciocalypta tyleri*.

TETRACTINELLIDA.

Tetilla hirsuta, n. sp. Tuticorin. Colour not recorded.

MONAXONIDA.

Halichondria panicea, *Johnston*, var. Rámésvaram. Light pink.
  Variety of the British species.
Petrosia testudinaria, *Lamarck*, sp. Tuticorin pearl banks. Pink, cup-shaped.

---

[1] Challenger Report on Monaxonida.
[2] A single small specimen of *Spongia officinalis* was collected by Dr. Anderson in the Mergui Archipelago.

Reniera madrepora, n. sp. Tuticorin. Colour not recorded.
Pachychalina multiformis, *Lendenfeld,* sp.(var. manaarensis nov.)
Tuticorin. Pale violet, or light pink.
,, delicatula, n. sp. Tuticorin. Colour not recorded.
,, spinilamella, n. sp. Tuticorin. Pale yellow.
Siphonochalina communis, *Carter*, sp. Tuticorin. Bluish brown.
Gelliodes carnosa, n. sp. Tuticorin. Grey.
Iotrochota baculifera, *Ridley* (var. flabellata, *Dendy*). Rámésvaram and Tuticorin. Dark purple.
Tedania digitata, *Schmidt*, sp. Rámésvaram. Red.
Clathria indica, n. sp. Tuticorin. Frequently incrusting pearl oyster. Bright red.
,, corallitincta, n. sp. Tuticorin. Coral-red.
Rhaphidophlus spiculosus, n. sp. Tuticorin. Vermilion.
Hymeniacidon ? fœtida, n. sp. Tuticorin. Grey ; smells like valerian when dry.
Phakellia ridleyi, n. sp. Rámésvaram. Red.
Ciocalypta tyleri, *Bowerbank* (var. manaarensis nov.). Tuticorin. White.
Acanthella carteri, n. sp. Tuticorin. Orange.
Auletta aurantiaca, n. sp. Tuticorin. Orange-red.
Axinella donnani, *Bowerbank.* Rámésvaram, and Tuticorin pearl banks. Orange.
,, labyrinthica, n. sp. Tuticorin. Orange.
,, tubulata, sp. *Bowerbank.* Rámésvaram, and Tuticorin pearl banks. Inhabited by commensal tubicolous Oligochœte worms. Pinkish-red or red.
Raspailia fruticosa, n. sp. Rámésvaram. Pink.
,, thurstoni, n. sp. Rámésvaram. Dry shore-specimens.
Suberites inconstans, n. sp. Between tide-mark's. Pámban.
,, *a* var. mœandrina. Brown. Canal system of var. mœandrina inhabited by Ophiuroids.
,, β var. digitata.
,, γ var. globosa.

CERATOSA.

Spongionella nigra, n. sp. Tuticorin. Black.
Hippospongia, sp. Rámésvaram.
Hircinia clathrata, *Carter.* Rámésvaram and Tuticorin. Canal system occupied by a cirrhiped crustacean, *Balanus (Acasta) spongites.*
,, vallata, n. sp. Rámésvaram.
Aplysina purpurea, *Carter.* Tuticorin. Grey (in spirit, or when dry) dark purple.
,, fusca, *Carter.* Tuticorin.

CÆLENTERATA.

OCTACTINIA.

Alcyonium digitulatum, *Klünz.* Rámésvaram.
„ gyrosum, *Klünz.* Rámésvaram.
„ polydactylum, *Ehr.* (var. mamillifera, *Klünz*). Rámésvaram.
Sarcophytum pauciflorum, *Ehr.*
Echinogorgia pseudosasappo, *Köll.* Rámésvaram; also from the Madras Harbour; studded, as figured by Esper, with Aviculæ and Ophiuroids.
„ sasappo, *Köll (Esper,* sp.). Rámésvaram.
„ cerea, *Köll.* Rámésvaram; also from the Madras Harbour.
„ furfuracea, *Köll (Esper,* sp.). Rámésvaram; also from the Madras Harbour.
Plexaura flabellum, *Esper.* Horny axes cast on shore at Rámésvaram and Tuticorin.
Juncella juncea, *Pallas.* Rámésvaram and Tuticorin (near shore and on pearl banks).
Gorgonia (Leptogorgia) miniacea, *M. Edw. (Esper,* sp.). Rámésvaram and Tuticorin.
Gorgonella umbella, *Esper.* Tuticorin.
Suberogorgia suberosa, *Pallas.* Rámésvaram and Tuticorin.
Corallium nobile, *Pallas.* Rámésvaram.
Pteroides javanicum, *Bleeker.* Rámésvaram.
„ esperi, *Herklots.* Rámésvaram and Tuticorin.
Virgularia juncea, *Esper.* Rámésvaram.
Lituaria, sp. Rámésvaram.

Some of the Alcyonia formed large, flat, encrusting masses. Entwining their arms round the stems and branches of *Juncella juncea, Suberogorgia suberosa,* etc., were delicate Ophiuroids (*Ophiothix,* etc.), and, clinging to the Gorgoniæ were the Crinoids, *Antedon reynaudi, Antedon palmata,* and *Actinometra parvicirra.* Living on the stems of the red-coloured Gorgoniæ I several times found the mollusc *Radius formosus,* the colour of whose shell corresponded with that of the Gorgoniæ.

A specimen of *Suberogorgia suberosa,* obtained at Mauritius in 90 fathoms, is described by Ridley (Ann. Mag. Nat. Hist., 1882, p. 132) as "an immense dry specimen 3 feet 5 inches high, and 18 inches in maximum lateral diameter. The colour is pale wainscot to pale rufous-brown; the branches are given off mostly at angles of 30°. The colour, very different from the deep brick-red usual in this species, may perhaps be due to the manner of drying." The height of a specimen in the Madras Museum from Tuticorin, where the pale and brick-red varieties were living side by side, is 4 feet 8 inches, and the maximum diameter 2 feet 2 inches.

The specimens of *Gorgonia miniacea* were characterised by the almost constant presence, on the stems or at their bifurcation, of diseased excrescences—the so-called galls— occupied by a Cirrhiped crustacean, and perforated by an orifice, through which currents of water for the respiration of the crustacean were admitted into the cavity of the excrescence, through which the stream passed in a constant direction. The association of similar excrescences on stony corals of the genera *Sideropora, Seriatopora*, and *Pocillopora*, is discussed in detail by Semper.[1]

Several fragments of *Corallum nobile*, the red coral of commerce, have been picked up by me on the Pámban beach, and the native divers tell me that they occasionally come across much larger pieces. Concerning this species Ridley says[2]:—

"Dr. Lankester (Uses of Animals to Man), besides the Persian Gulf, gives Ceylon as a locality for this, the precious red coral of the Mediterranean and Cape Verd Islands, and Dr. Ondaatje has shown me decorticated specimens from Ceylon, which make the identity of the species probable. It is noteworthy that a fossil form is recorded from Indian deposits (Duncan), which as I have given reasons for thinking (Proc. Zool. Soc., 1882, p. 334) seems probably identified with this species, Seguenza having found it fossil in India, still bearing a slight red tint. 'An officer,' in a work entitled 'Ceylon' (London, 8vo., 1876) mentions small fragments of red coral similar to that of the Mediterranean as having been found at the water's edge between Galle and Colombo, and states it to have been referred to by the Portuguese."

It must be borne in mind, however, that the red coral of commerce is imported to the East in large quantities to be worked up into ornaments for natives; and it is possible that the small fragments, picked up from time to time on the beach, may be only adventitious products, and not a natural product of the neighbouring sea.

---

[1] "Animal Life." *Internat. Science Ser.*, vol. XXXI.
[2] *Ann. Mag. Nat. Hist.*, vol. XI, 1883.

The sea-pen *Virgularia juncea* was collected at low water, and accords in its habits with another species, *V. patagonica*, which is described by Darwin [1] as being seen projecting like stubble, with the truncate end upwards, a few inches above the surface of the muddy sand. When touched or pulled, they suddenly drew themselves in with force, so as nearly or quite to disappear.

HEXACTINIA.

ACTINIARIA.

Various undetermined species of Sea-anemone are found, either burrowing in the sandy shore between tide-marks, attached to, or living within cavities excavated in coral blocks. A single specimen of *Palythoa tuberculosa*, recorded by Esper from Tranquebar on the East Coast of the Madras Presidency, was brought up by the divers at Pámban, encrusting the upper surface of a dead coral. Various species of *Zoanthus*, single or colonial, live among the corals on the reefs. At both Tuticorin and Pámban I have several times seen specimens of *Sphenopus marsupialis*, which was collected originally by Johns, a Moravian Missionary, at Tranquebar, and was cast on shore at Madras in large numbers during the cyclone of 1886. The outer surface of this species is made up of sand grains glued together by a viscid secretion, and imbedded in a cartilaginous case. Specimens figured in the Proceedings of the Zoological Society, Feb. 14, 1867, were collected at Pulo Faya in the China seas.

MADREPORARIA.

(Stony Corals.)

Madrepora cervicornis, *Lamk.*
,, cytharea, *Dana.*
,, spicifera, *Dana.*
Montipora digitata, *Dana.*
,, foliosa, *Pallas.*
,, irregularis, *Quelch.*
,, tuberosa, *Klünz.*
Porites lutea, (*Q. and Gaim*), *M. Edw. and H.*
,, solida, *Forskal.*
Turbinaria cinerascens, *Solander*, sp.
,, crater.
,, patula, *Dana.*

---

[1] *Journal of Researches.*

Cænopsammia ehrenbergiana, *M. Edw. and H.*
Pocillopora damicornis, *Esper.*
,,   hemprichii, *Ehr.*
,,   favosa, *Ehr.*
,,   verrucosa, *Ell. Sol.*
Galaxea bougainvillei, *M. Edw. and H.*
,,   irregularis, *M. Edw. and H.*
Mussa radians.
Leptoria tenuis, *Dana.*
Cæloria arabica, *Klünz.*
,,   ,,   var. subdentata.
Hydnophora conicolobata, *M. Edw. and H.*
,,   lobata, *Lamk.*
,,   microconus, *Lamk.*
Favia ehrenbergi, *Klünz*, var.
,,   aspera, *M. Edw. and H.*
Goniastræa (undetermined species).
Prionastræa (undetermined species).
Cyphastræa savignii, *M. Edw. and H.*
Echinopora lamellosa, *Esper.*
Fungia, sp., juv.
Cycloseris cyclolites, *Lamk.*
Heteropsammia cochlea, *Spengler.*
Pavonia, sp.
Psammocora acerosa, *Brügemann.*

The majority of these stony corals belong to the class of 'reef corals,' but a few species are included, *e.g.*, *Cycloseris cyclolites* and *Heteropsammia cochlea*, which were dredged in deep water, where the reef-builders were absent, and the young *Fungiæ*, which were dredged from the muddy bottom of the Pámban Pass. All the specimens of *H. cochlea* exhibited a hole bored by a sipunculid worm (*Aspidosiphon*), which is always found living within this coral. It is difficult, as Semper points out,[1] to understand what advantage each animal can derive from their association; yet some must exist, for a coral is never found without a worm.

The fact is drawn attention to by Dr. Martin-Duncan, in his report[2] on the Madreporaria of the Mergui Archipelago collected by Dr. Anderson, as being very remarkable that "the Coral-fauna of Ceylon, so far as it is known from Mr. Stuart O. Ridley's researches,[3] does not contain a single Mergui species. The number of genera common to the two areas is, however, great, and many species are closely allied."

---

[1] "Animal Life." *Internat. Science Ser.*, 1881.
[2] *Journ. Linn. Soc.*, Nov. 13, 1886.
[3] *Ann. Mag. Nat. Hist.*, 1883.

A comparison of the list of species recorded above from the Indian side of the Gulf of Manaar with those of Dr. Duncan and Mr. Ridley shows, as might be expected, that some of the species are common to the Indian coast of the Gulf of Manaar and Ceylon, and others to the Indian coast of the Gulf of Manaar and the Mergui Archipelago.

The type specimens of *Montipora irregularis*, of which species I obtained several specimens from the Pámban reef, were obtained during the voyage of the Challenger at Samboangan, Philippines.

I have found no representative of the Hydrocorallinæ on the coral reefs, but *Millepora dichotoma* has been recorded by Ridley [1] from Ceylon.

The genus *Heliopora* is apparently not represented on the living reef, but a single specimen of *Heliopora edwardsana* has been described from the cretaceous deposits of the Trichinopoly District of the Madras Presidency, concerning the coral-beds of which Stoliczka writes [2]:—

"The conditions of the deposits were not so quiet that we could expect to find any of the Alcyonaria or of the Malacodermata preserved, but the Sclerodermata or Madreporaria are represented by fifty-seven species, namely, fifty-three belonging to the *Aporosa*, three to the *Perforata*, and one to the *Tabulata* ......Looking at the whole fauna we see the reef-building *Astræidæ*, *Stylinidæ*, and *Thamnastræidæ* much exceeding the other families in numbers of species, as well as in frequency of occurrence of specimens. Coral reefs appear to have been of considerable extent, particularly along the old shores within the Ootatoor group; in the two other groups they were much more local."

The method employed by me for the preservation of corals (*i.e.*, the skeletons) which I reserve for exhibition, is to expose them to the action of the sun and ants, which remove a large amount of the animal matter, and send them in boxes, surrounded by paper and tightly packed in rice-husk, by native sailing boat to Madras. But, however great the care which is taken, it generally happens that some of the corals become covered with mould during the voyage. The rice-husk, which is usually found clinging to the surface of the corals, is removed with a syringe, and the corals, after being submitted to repeated washings with fresh-water, are

---

[1] *Ann. Mag. Nat. Hist.* Ser. 5, vol. XI, 1883.
[2] *Palæont Ind. Cretaceous Fauna of Southern India.*

finally dried in the sun. In no case are they submitted to the action of corrosive alkali solutions. It has been objected, with regard to the preservation of corals by exposing them for sometime to the action of rain or running water, that the finest details of the skeleton are liable to be dissolved away to some extent by the action of the carbonic acid in the water. But I found, on my visit to Rámésvaram Island in 1889, that the structural details of various delicate corals (*Astræopora, Cyphastræa*, etc.), which I had left discarded on the sand in the " compound " of the bungalow twelve months previously, were, to no appreciable extent, damaged for purposes of identification, though they had, in the interval, been freely exposed to the action of a heavy monsoon and a cyclone.

ECHINODERMATA.

A report on a collection of Echinoderms, which I made in the years 1886-87 at Rámésvaram island and Tuticorin, by Prof. F. Jeffrey Bell, has been published in the Proceedings of the Zoological Society, June 19, 1888, wherein the writer states that " I may be allowed to remind the student of the recent appearance of a memoir on the Echinoderm-fauna of the island of Ceylon.[1] Shortly after the distribution of that memoir my respected correspondent, M. de Loriol, was kind enough to write and tell me of four other species of Echinoids, all of which had been collected at Aripo by M. Alois Humbert." Of these four species (*Phyllacanthus annulifera, Temnopleurus reynaudi, Clypeaster humilis*, and *Laganum depressum*) *C. humilis* and *L. depressum* have been found by me off the Indian coast of the Gulf of Manaar.

Only two new species have been discovered among my collections, viz., an Ophiuroid, *Pectinura intermedia*, and an Asteroid, *Oreaster (Pentaceros) thurstoni*, of which the latter is a very common shallow-water species very variable both in its characters and colour. Since the publication of Prof. Bell's Report several species, not recorded there, have been found in my subsequent visits to the Gulf of Manaar, bringing the total number up to fifty-eight.

The majority of the specimens were found in shallow-water near the shore, but some, *e.g., Oreaster (Pentaceros) lincki, Linckia lævigata, Colochirus quadrangularis*, and *Astrophyton*, sp. (of which a single imperfect specimen was found

---

[1] *Scientific Transactions of the Royal Dublin Society* (2), III, p. 643 *et seq.*

within the cup of a *Turbinaria*) were brought up by divers from the pearlbanks in ten to eleven fathoms.

Of the six species of Echinoid which are described by Agassiz, in his 'Revision of the Echini' as being characteristic of his Indo-African Region, which includes the Madras coast, five, viz., *Echinodiscus auritus* and *biforis*, *Salmacis sulcata* and *bicolor*, and *Echinolampus oviformis*, are very abundant in the Gulf of Manaar. But I have not as yet found the sixth species, *Echinodiscus lævis*.

The fossil Echinodermata, as recorded in the 'Palæontologia indica' from the cretaceous deposits in South India, are represented by two or three species of Crinoidea (*Pentacrinus* and *Marsupites*), a single species of Asteroidea (*Ophiura ? cunliffei*), and thirty-eight species of Echinoidea, of which the genera *Cidaris* and *Hemiaster* are most largely represented.

CRINOIDEA.

Antedon palmata, *Müll.* Rámésvaram and Tuticorin. In crevices of coral or on *Gorgoniæ*.
,, reynaudi, *Müll.* Rámésvaram. On stems of *Gorgoniæ*.
Actinometra parvicirra, *Müll.* Tuticorin. On stems of *Gorgoniæ*.

ASTEROIDEA.

Echinaster purpureus (*Gray*), *Bell.* Tuticorin.
Linckia lœvigata,[1] *Gmelin.*[1] Pearl banks.
Anthenea acuta, *Perrier.* Rámésvaram.
,, pentagonula (*Lamk.*), *Perrier.* Rámésvaram.
Goniodiscus granuliferus (*Gray*), *Perrier.* Rámésvaram.
Stellaster, sp. ? Pearl banks.
Oreaster lincki, *de Bl.*[2] Pearl banks.
,, superbus, *Möbius.*[3] Tuticorin.
,, thurstoni, n. sp. *Bell.*[4] Rámésvaram and Tuticorin.
Asterina cepheus (*M. Tr.*), *Von. Mart.* Rámésvaram.
Luidia hardwickii (*Gray*), *Perrier.* Rámésvaram.
,, maculata, *M. Tr.* Rámésvaram.
,, ,, sp.? Rámésvaram.
Astropecten hemprichii, *M. Tr.* Rámésvaram. A specimen in the Madras Museum has swallowed a mollusc, *Cerithium* sp.

---

[1] *L. miliaris* (Linck), Von Martens, and *L. lævigata*, Lütken. Vide Challenger Report, Asteroiden.
[2] *Pentaceros muricatus*, Linck. Challenger Report.
[3] *Pentaceros superbus*, Möbius, sp. Challenger Report.
[4] *Pentaceros thurstoni*, Bell, sp. Challenger Report.

Astropecten polyacanthus, *M. Tr.* Rámésvaram.
„   „   sp., juv. Rámésvaram.

OPHUIROIDEA.

I am indebted to Professor Jeffrey Bell for the identification of the following additional Echinoderms, collected by me at Tuticorin during the pearl fishery of 1889 :—

CRINOIDEA.

Antedon cumingi.

ASTEROIDEA.

Asterodiscus elegans, *Gray.*
Pentaceros, n. sp. ?

OPHIUROIDEA.

Pectinura infernalis. *Ltk.*
Ophionereis dubia, *Lym.*

ECHINOIDEA.

Toxopneustes pileolus, *Ag.*
Pseudoboletia maculata.
Phyllacanthus baculosa, *A. Ag.*

Lovenia elongata, *And.* Rámésvaram.

HOLOTHUROIDEA.

Cucumaria semperi, *Bell.* Rámésvaram.
Colochirus quadrangularis, *Less.* Pearl banks.
Actinocucumis difficilis, *Bell.* Rámésvaram.

---

[1] Prof. E. V. Martens, to whom a specimen was sent, reports that he cannot find any real difference between it and Müller's type of *Ophiothrix aspidota*, with which he compared it.

within the cup of a *Turbinaria*) were brought up by divers from the pearlbanks in ten to eleven fathoms.

Of the six species of Echinoid which are described by Agassiz, in his ' Revision of the Echini ' as being character-

,, maculata, *M. Tr.* Rámésvaram.
,, ,, sp.? Rámésvaram.
Astropecten hemprichii, *M. Tr.* Rámésvaram. A specimen in the Madras Museum has swallowed a mollusc, *Cerithium* sp.

---

[1] *L. miliaris* (Linck), Von Martens, and *L. lœvigata*, Lütken. *Vide* Challenger Report, Asteroidea.
[2] *Pentaceros muricatus*, Linck. Challenger Report.
[3] *Pentaceros superbus*, Möbius, sp. Challenger Report.
[4] *Pentaceros thurstoni*, Bell, sp. Challenger Report.

Astropecten polyacanthus, *M. Tr.* Rámésvaram.
,,   ,,   sp., juv. Rámésvaram.

OPHUIROIDEA.

Pectinura gorgonia, *M. Tr.* Rámésvaram.
,,   intermedia, n. sp., *Bell.* Rámésvaram.
Ophiactis savignii, *Audouin.* Rámésvaram. In canal system of a sponge, *Suberites inconstans.*
Ophiocoma erinaceus, *M. Tr.* Rámésvaram.
Ophiothrix longipeda, *M. Tr.* Rámésvaram.
,,   nereidina, *M. Tr.* Rámésvaram.
,,   aspidota,[1] *M. Tr.* Rámésvaram.
Ophiomaza cacaotica, *Lyman.* Rámésvaram.
Hemieuryalid. Rámésvaram. Arms entwined round branches of Suberogorgia suberosa.
Astrophyton clavatum, *Lyman?* Pearl banks.

ECHINOIDEA.

Temnopleurus toreumaticus, *Leske.* Rámésvaram.
Temnopleuroid.
Salmacis bicolor, *Ag.* Pearl banks.
,,   dussumieri, *Ag.* Rámésvaram. Common in fishing-nets at Madras.
,,   sulcata, *Ag.* Tuticorin.
Stomopneustes variolaris, *Lamk.* Rámésvaram.
Echinometra lucunter, *Leske.* Tuticorin.
Fibularia volva, *Ag.* Rámésvaram.
Clypeaster humilis, *Leske.* Pearl banks.
Laganum decagonale, *Less.* Rámésvaram.
,,   depressum, *Less.* Rámésvaram.
Echinodiscus biforis, *Ag.* Tuticorin.
,,   auritus, *Leske.* Rámésvaram.
Echinolampas oviformis, *Gmelin.* Rámésvaram.
Rhinobrissus pyramidalis, *A. Ag.* Rámésvaram.
Brissus unicolor, *Leske.* Rámésvaram.
Metalia sternalis, *Lamk.* Tuticorin.
Lovenia elongata, *And.* Rámésvaram.

HOLOTHUROIDEA.

Cucumaria semperi, *Bell.* Rámésvaram.
Colochirus quadrangularis, *Less.* Pearl banks.
Actinocucumis difficilis, *Bell.* Rámésvaram.

---

[1] Prof. E. V. Martens, to whom a specimen was sent, reports that he cannot find any real difference between it and Müller's type of *Ophiothrix aspidota*, with which he compared it.

Haplodactyla australis, *Semper.* Tuticorin.
Holothuria atra, *Jäger.* Rámésvaram.
,,   marmorata, *Jäger.* Rámésvaram. Edible Trepang.
,,   monacaria, *Less.* Rámésvaram.
,,   vagabunda, *Selenka.* Tuticorin.
Synapta recta, *Semper?* Rámésvaram.
Thyone sacella, *Selenka.* Rámésvaram. A specimen in the Madras Museum shows the tentacles, teeth, etc., which were ejected during life.

CRUSTACEA.

(Decapoda and Stomatopoda.)

As regards the Crustacea (Decapoda and Stomatopoda) Mr. J. R. Henderson writes to me:

"This collection is one of the most important which has ever been formed on the Indian coast. It contains over a hundred and twenty species, not more than ten or twelve of which are new to science; but a number of rare or little-known forms are present, and the geographical distribution of most of these has been greatly extended by their discovery on the South Indian shores. Upwards of three hundred species of Decapod and Stomatopod Crustacea have been recorded from the Bay of Bengal, which may be conveniently held to include the coasts from Ceylon on the one side to Singapore on the other, along with the numerous groups of islands situated within this area. Yet, with the exception of a small collection from Madras report on by Prof. Heller in the Crustacea of the 'Reise der Novara,' our knowledge of the species which inhabit the Indian coast proper is limited to a few scattered papers, and to those forms recorded by the older writers under the somewhat vague localisation 'Indian Seas.'

"The Crustacean fauna of the Gulf of Manaar shows, as might be expected, a considerable proportion of coral reef species—widely distributed forms, which occur in suitable localities throughout the vast Indo-Pacific region. The following list does not include the new species, or a few which are still unidentified. Those new to the Bay of Bengal (as defined above) are marked with an asterisk."

DECAPODA.

BRACHYURA.

Oxyrhyncha.

Doclea hybrida (*Fabr.*).
Hyastenus hilgendorfi, *DeMan.*

* Stenocionops cervicornis (*Herbst*).
* Huenia proteus, *DeHaan*.
* Menæthius monoceros (*Latr.*).
* Zebrida adamsii, *White*.
* Tylocarcinus styx (*Herbst*).
* Micippe philyra (*Herbst*).
* ,, thalia (*Herbst*).
* Lambrus affinis, *A. M. Edw.*
  ,, longispinus, *Miers.*
  ,, holdsworthi, *Miers.*
* Paratymolus sexspinosus, *Miers.*

Cyclometopa.

Atergatis integerrimus (*Lam.*).
,, floridus (*Rumph.*).
* Carpilodes margaritatus, *A. M. Edw.*
* Lophactæa granulosa (*Rüppell*).
* Actæa nodulosa (*White*).
  ,, ruppellii (*Krauss*).
* ,, granulata (*Aud.*).
* Euxanthus melissa (*Herbst*).
* Xantho punctatus, *M. Edw.*
* Lophozozymus dodone (*Herbst*).
* Cycloxanthus lineatus, *A. M. Edw.*
  Chlorodius niger (*Forskall*).
* Phymodius monticulosus (*Dana*).
  Leptodius exaratus (*M. Edw.*)
  Etisus lævimanus (*Randall*).
* Etisodes sculptilis, *Heller.*
  Cymo andreossyi (*Aud.*).
  Menippe rumphii (*Fabr.*).
  Pilumnus vespertilio (*Fabr.*).
* ,, longicornis, *Hilg.*
  Eriphia lævimana, *Guérin.*
  Trapezia rufopunctata (*Herbst*).
  ,, cymodoce (*Herbst*).
  ,, areolata, *Dana.*
* Tetralia nigrifrons, *Dana.*
  Neptunus pelagicus (*Linn.*).
  ,, sanguinolentus (*Herbst*).
  ,, gladiator (*Fabr.*).
  Scylla serrata (*Forskal*).
  Thalamita prymna (*Herbst*).
  ,, admete (*Herbst*).
  ,, savignyi, *A. M. Edw.*
  Goniosoma natator (*Herbst*).
  ,, annulatum (*Fabr.*).
* ,, callianassa (*Herbst*).

Goniosoma cruciferum (*Fabr.*).
* ,, orientale (*Dana*).
,, merguiense, *DeMan*.
* Lissocarcinus lœvis, *Miers*.

Catometopa.

Telphusa corrugata, *Heller*.
Cardisoma carnifex (*Herbst*).
Ocypoda ceratophthalma (*Pallas*).
,, cordimana, *Desm*.
,, platytarsis, *M. Edw*.
,, macrocera, *M. Edw*.
Gelasimus annulipes, *Latr*.
* Xenophthalmus pinnotheroides, *White*.
Metopograpsus messor (*Förskal*).
Grapsus strigosus (*Herbst*).
,, maculatus (*Catesby*).
Leiolophus planissimus (*Herbst*).
Sesarma quadrata (*Fabr.*).

Oxystomata.

Calappa hepatica (*Linn.*).
,, gallus (*Herbst*).
Matuta victrix (*Fabr.*).
,, lunaris (*Herbst*).
,, miersii, *Henderson*.
Leucosia craniolaris (*Linn.*).
Philyra scabriuscula (*Fabr.*).
,, globosa (*Fabr.*).
* Myra fugax (*Fabr.*).
Dorippe fachino (*Herbst*).
,, quadridentata (*Fabr.*).

Anomura

Dromidia unidentata (*Rüppell*).
Hippa asiatica, *M. Edw*.
Albunea symnista (*Fabr.*).
Diogenes miles (*Fabr.*).
,, custos (*Fabr.*).
* ,, diaphanus (*Fabr.*).
,, avarus, *Heller*.
Pagurus punctulatus. *Olivier*.
,, deformis, *M. Edw*.
Clibanarius padavensis, *DeMan*
* Aniculus typicus, *Dana*.
Cœnobita rugosa, *M. Edw*.
Petrolisthes boscii (*Aud.*).

Petrolisthes dentatus (*M. Edw.*).
* Porcellanella triloba, *White.*
Polyonyx obesulus (*White*).
Galathea elegans, *White.*

MACRURA.

Palinurus dasypus, *Latr.*
Gebiopsis intermedia, *DeMan.*
Alpheus edwardsii (*Aud.*).
,, lævis, *Randall.*
,, comatulorum, *Haswell.*
* Betæus frontalis, *M. Edw.*
* Pontonia tridacnæ, *Dana.*

STOMATOPODA.

* Lysiosquilla maculata (*Fabr.*).
Squilla scorpio, *Latr.*
,, nepa, *Latr.*
* Pseudosquilla ciliata (*Fabr.*).
Gonodactylus graphurus, *White.*
* ,, trispinosus, *White.*

MOLLUSCA.

The following list of Marine mollusca, which I have collected in the Gulf of Manaar, includes (1) those which were collected on the beach, all shells which were worn and bore evidence of having been rolled in from a distance being rejected, and only those which appeared to be fresh being retained; (2) those which were obtained by dredging, and straining the contents of the dredge through sieves; (3) those which were collected on the coral reefs on clear days or at low tide; (4) those which were brought up from the pearl banks and other places by native divers; (5) those which were obtained by examining the sweepings from the kottus during the pearl fishery; (6) those which were found attached to *Algæ* and *Gorgoniæ*, or obtained by breaking up coral blocks with a crowbar, and extracting the shells which were buried in holes bored by the animals during life.

CEPHALOPODA.

Spirula peronii, *Lmk.* Rámésvaram.
Nautilus pompilius, *Linn.* Rámésvaram.

PTEROPODA.

Styliòla acicula. Pelagic over Pámban reef.

HETEROPODA.

Ianthina africana, *Reeve.* Very abundant, coincidently with *Physalia*, on the surface, one evening at Kílakarai.

GASTEROPODA.

Murex anguliferus, *Lmk.* Tuticorin.
,, haustellum, *Linn.* Rámésvaram and Tuticorin.
,, tenuispina, *Lmk.* Tuticorin.
,, (Ocinebra) sp. Rámésvaram.
Pleurotoma tigrina, *Lmk.* Rámésvaram.
,, (Surcula) javana, *deBoiss.* Rámésvaram.
,, (Drillia) inconstans, *Smith.* Rámésvaram.
Fusus colus, *Linn.* Rámésvaram.
,, tuberculatus, *Lmk.* Rámésvaram.
Cythara pallida, *Reeve.* Rámésvaram.
Tritonidea undosa, *Linn.* Rámésvaram.
Triton chlorostomus, *Lmk.* Rámésvaram.
,, cingulatus, *Pfeiff.* Tuticorin.
,, retusus, *Lmk.* Tuticorin.
,, (Persona) cancellinus, *deBoiss.* Tuticorin.
Ranella foliata, *Brod.* Tuticorin.
Ranella granifera, *Lmk.* Rámésvaram.
Buccinum (Cantharus) melanostoma, *Sow.* Tuticorin.
Cyllene, sp. Rámésvaram.
Nassa callospira, Rámésvaram.
,, ornata, *Kien.* Rámésvaram.
,, suturalis, *Lmk.* Rámésvaram.
,, sp. juv. Rámésvaram.
,, (Arcularia) bimaculosa, *A. Ad.* Rámésvaram.
,, (Niotha) albescens, *Dunk.* Rámésvaram.
,, ,, australis, *A.Ad.* Rámésvaram.
Eburna spirata, *Lmk.* Rámésvaram.
,, zeylanica, *Lmk.* Rámésvaram.
Purpura rudolphi, *Lmk.* Rámésvaram.
Oliva candida, *Lmk.* Rámésvaram.
,, gibbosa, *Born.* Rámésvaram.
,, ispidula, *Linn.* Rámésvaram.
Ancillaria, sp. Rámésvaram.
,, (Sparella) cinnamonea, *Lmk.* Tuticorin.
,, ,, ampla, *Gmel.* Tuticorin.
Fasciolaria filamentosa, *Chemn.* Rámésvaram.
Tudicla spirillus, *Lmk.* Rámésvaram.
Turbinella cornigera, *Lmk.* Tuticorin.
,, pyrum, *Lmk.* Tuticorin.

Turbinella rapa, *Lmk.* Tuticorin. The chank.
Voluta interpuncta, *Martyn.* Tuticorin.
Cymbium indicum, *Gmel.* Rámésvaram.
Marginella angustata, *Sow.* Rámésvaram and Tuticorin.
,, dens, juv. *Reeve.* Tuticorin.
,, n. sp.? Rámésvaram.
Zafra atrata, *Gould.* Rámésvaram.
Engina zonata, *Reeve.* Rámésvaram.
,, sp. juv. Rámésvaram.
Harpa ventricosa, *Lmk.* Rámésvaram.
Cassis areola, *Lmk.* Rámésvaram.
,, canaliculata, *Lmk.* Rámésvaram.
,, (Bezoardica) glauca, *Brug.* Rámésvaram.
Dolium olearium, *Linn.* Rámésvaram.
,, fasciatum, *Lmk.* Rámésvaram.
,, maculatum, *Lmk.* Rámésvaram.
Ficula reticulata, *Lmk.* Rámésvaram.
Pyrula (Melongena) vespertilio, *Lmk.* Rámésvaram.
Natica maroccana, *Chemn.* Rámésvaram.
,, pulicaria, *Phil.* Tuticorin.
,, sp. juv. Tuticorin.
,, (Mamilla) melanostoma, *Lmk.* Tuticorin.
,, (Neverita) didyma, *Bolt.* Rámésvaram.
,, (Polinices) mamilla, *Lmk.* Rámésvaram.
Sigaretus neritoideus, *Linn.* Rámésvaram.
Naticina, sp. juv. Rámésvaram.
Scalaria aculeata, juv. *Sow.* Rámésvaram.
,, decussata, *Pease.* Rámésvaram.
,, sp. Rámésvaram.
Terebra myuros, *Lmk.* Rámésvaram.
,, n. sp.? Rámésvaram.
Ringicula doliaris? *Gould.* Tuticorin.
,, sp. Tuticorin.
Chemnitzia, sp. Tuticorin.
Alaba rectangularis, *Cramer.* Rámésvaram.
Solarium perspectivum, *Lmk.* Rámésvaram.
,, lævigatum, *Lmk.* Rámésvaram.
,, (Torinia) cœlata, *Hinds.* Rámésvaram.
,, ,, fulvum, *Hinds.* Rámésvaram.
Conus amadis, *Martini.* Rámésvaram.
,, figulinus, *Linn.* Tuticorin.
,, hebræus, *Linn.* Rámésvaram.
,, marmoreus, *Linn.* Rámésvaram.
,, piperatus, *Dillwyn.* Rámésvaram.
,, striatus, *Linn.* Tuticorin.
,, textile, *Linn.* Tuticorin.
Strombus canarium, *Linn.* Rámésvaram and Tuticorin.
Pterocera aurantia, *Lmk.* Rámésvaram.
,, lambis, *Linn.* Rámésvaram.

Pterocera scorpius, *Linn.* Rámésvaram.
Cypræa arabica, *Linn.* Rámésvaram and Tuticorin.
,, ,, var. Tuticorin.
,, caput-serpentis, *Linn.* Tuticorin.
,, carneola, *Linn.* Tuticorin.
,, errones, *Linn.* Rámésvaram.
,, hirundo, *Gmel.* Rámésvaram.
,, lynx, *Linn.* Rámésvaram.
,, mauritiana, *Linn.* Rámésvaram and Tuticorin.
,, moneta, *Linn.* Rámésvaram. The money cowry.
,, ocellata, *Linn.* Rámésvaram and Tuticorin.
,, talpa, *Linn.* Tuticorin.
,, tigris, *Linn.* Rámésvaram and Tuticorin.
,, vitellus, *Linn.* Tuticorin.
Ovulum (Radius) birostre, *Linn.* Rámésvaram.
,, ,, formosus, *Ad. and Reeve.* Rámésvaram.
Attached to branches of red *Gorgoniæ*.
Ovulum (Radius) volva, *Linn.* Rámésvaram.
Cerithium breviculum, *Sow.* Rámésvaram.
,, morus, *Lmk.* Rámésvaram.
,, (Tympanotomus) alatum, *Phil.* Rámésvaram.
,, ,, fluviatile, *Potiez.* Rámésvaram.
Triforis cingulatus, *Ad.* Tuticorin.
,, corrugatus, *Hinds.* Rámésvaram.
,, ruber, *Hinds.* Rámésvaram.
Pyrazus palustris, *Linn.* Rámésvaram.
Cerithidea, sp. Rámésvaram.
Melania collistricta, *Reeve.* Tuticorin.
Littorina scabra, *Linn.* Rámésvaram and Tuticorin.
Rissoina planaxoides. Rámésvaram.
Turritella attenuata, *Reeve.* Rámésvaram.
Siliquaria encaustica, *Mörch.* Rámésvaram and Tuticorin.
Calyptra layardi, *Reeve.* Rámésvaram.
Galerus extinctorium, *Sow.* Rámésvaram.
Crepidula walshi, *Hermanns.* Rámésvaram and Tuticorin.
Capulus, sp. Rámésvaram.
Vanicora quoyiana, *A. Ad.* Rámésvaram.
Nerita plicata, *Linn.* Rámésvaram.
,, rumphii, *Recluz.* Rámésvaram.
,, squamulata, *LeGuill.* Rámésvaram.
Neritina, sp. Rámésvaram.
Clypeola, sp. Rámésvaram and Tuticorin.
Phasianella lineolata, *Wood.* Rámésvaram.
Turbo (Senectus) margaritaceus, *Linn.* Rámésvaram.
,, (Collonia, sp. Rámésvaram.
Liotia cidaris, *Reeve.* Rámésvaram.
Rotella costata, *Valenc.* Rámésvaram.
,, vestiaria, *Sow.* Tuticorin.
Delphinula distorta, *Lmk.* Rámésvaram and Tuticorin.

Trochus niloticus, *Linn.* Rámésvaram.
,, (Eutrochus), sp. Rámésvaram.
,, (Euchelus) atratus, *Gmel.* Rámésvaram.
,, (Clanculus) microdon. *A. Ad.* Rámésvaram.
,, (Zizyphinus) tranquebaricus, *Chemn.* Rámésvaram.
,, (Monilea) solandri, *Phil.* Rámésvaram.
,, (Solariella), sp. Rámésvaram.
,, (Gibbula) variabilis, *Ad.* Tuticorin.
Haliotis semistriata, *Reeve.* Rámésvaram.
,, varia, *Linn.* Rámésvaram.
Fissurella singaporensis, *Reeve.* Tuticorin.
,, (Lucapina) quadriradiata, *Reeve.* Rámésvaram.
Marginula, sp. Rámésvaram.
Scutum unguis, *Linn.* Rámésvaram and Tuticorin.
Dentalium octogonum, *Lmk.* Tuticorin.
Scutellina asperulata, *A. Ad.* Rámésvaram.
Chiton, several species. Rámésvaram.
Solidula solidula, juv. *Lmk.* Rámésvaram.
Hydatina circulata, *Martyn.* Rámésvaram.
Bulla ampulla, *Linn.* Rámésvaram.
Haminea cymbalum, *Quoy and Gaim.* Rámésvaram.
Atys porcellana, juv. *Gould.* Rámésvaram.
Lobiger viridis,[1] *G. and H. Nevill.* Tuticorin.
Dolabella, sp. Rámésvaram and Tuticorin.
Siphonaria, sp. Rámésvaram.

LAMELLIBRANCHIATA.

Dactylus orientalis, *Gmel.* Rámésvaram.
Jouannetia globosa, *Quoy.* Rámésvaram. From cavities bored in coral.
Parapholas, sp. Rámésvaram. From cavities bored in wood.
Guetra nucifera, *Spengl.* From cavities bored in wood.
Rocellaria gigantea, *Deshay.* Rámésvaram. From cavities in coral.
Brechites, sp. Tuticorin.
Solen adspersus, juv. *Dunk.* Rámésvaram
,, corneus, *Lmk.* Tuticorin.
,, linearis, juv? *Chemn.* Rámésvaram.
Siliqua radiata, *Linn.* Rámésvaram.
Corbula fortisulcata, *Smith.* Tuticorin.
,, modesta, *Hinds.* Rámésvaram.
,, sulculosa, *H. Ad.* Rámésvaram.
Theora fragilis, *A. Ad.* Rámésvaram and Tuticorin.
Soletellina diphos, *Linn.* Rámésvaram.
Tellina, sp. juv. Rámésvaram.
,, (Arcopagia) pristis, *Lmk.* Rámésvaram.

---

[1] Vide *Ann. Mag., Nat. Hist.*, 1889.

Dosinia, sp. juv. Rámésvaram.
Donax cuneata, *Linn.* Tuticorin.
,, faba, *Chemn.* Rámésvaram.
,, scortum, *Linn.* Rámésvaram.
Semele exarata, *Ad. and Reeve.* Rámésvaram.
Paphia trigona, *Deshay.* Tuticorin.
Ervillia, sp. Rámésvaram.
Venus lamarckii, *Gray.* Rámésvaram.
,, lamellaris, *Schumach.* Rámésvaram.
,, reticulata, *Linn.* Rámésvaram.
,, (Sunetta) meroe, *Linn.* Rámésvaram.
,, ,, scripta, *Linn.* Rámésvaram.
,, ,, truncata, *Deshay.* Rámésvaram.
,, (Chione) layardi, *Sow.* Rámésvaram.
,, ,, scabra, *Hanley.* Rámésvaram.
Cytherea (Callista) erycina, *Linn.* Rámésvaram.
,, (Circe) personata, *Deshay.* Rámésvaram.
,, (Meretrix) casta, *Hanley.* Rámésvaram.
Tapes adspersa, *Chemn.* Rámésvaram.
,, litterata, *Linn.* Rámésvaram.
,, rotundata, *Linn.* Rámésvaram.
,, textrix, *Chemn.* Rámésvaram.
,, undulata, *Born.* Tuticorin.
,, malabarica, *Chemn.* Tuticorin.
Venerupis macrophylla, *Deshay.* Rámésvaram. From cavities bored in coral.
Petricola lapicidum, *Chemn.* Rámésvaram. From cavities bored in coral.
Cardium asiaticum, *Brug.* Tuticorin.
,, rugosum, *Lmk.* Tuticorin.
,, rubicundum, *Reeve.* Rámésvaram.
Lævicardium australe, *Sow.* Rámésvaram.
,, retusum, *Linn.* Rámésvaram.
Lunulicardia subretusa, *Sow.* Rámésvaram.
Lucina (Divaricella) cumingii, *Ad. and Ang* Tuticorin.
Codakia fischeriana, *Issel.* Rámésvaram.
,, sp. Rámésvaram.
Cryptodon vesicula, *Gould.* Tuticorin.
Diplodonta, sp. Rámésvaram.
Scintilla, sp. Rámésvaram.
Crassatella radiata, *Sow.* Rámésvaram and Tuticorin.
Cardita antiquata, *Reeve.* Tuticorin.
,, canaliculata, juv. *Reeve.* Rámésvaram.
Modiolaria. Rámésvaram.
Modiola japonica, juv. *Dunk.* Rámésvaram.
,, tulipa, *Lmk.* Rámésvaram.
Lithodomus stramineus, *Dunk.* Tuticorin. From cavities bored in coral.

Lithodomus teres, *Philipp.* Rámésvaram and Tuticorin. From cavities bored in coral.
Avicula fucata, *Gould.* Tuticorin. The pearl oyster.
,, inquinata, *Reeve.* Rámésvaram. On the branches of *Gorgoniæ.*
Avicula radiata, *Pease.* Tuticorin. On the branches of *Gorgoniæ.*
Avicula vexillum, *Reeve.* Rámésvaram and Tuticorin. Sometimes found in masses attached to the branches of hyroids.
Arca sculptilis, *Reeve.* Rámésvaram. From cavities in coral.
,, virescens, *Reeve.* Rámésvaram. From cavities in coral.
,, (Anadara) granosa, *Linn.* Rámésvaram.
,, (Barbatia) lima, *Reeve.* Tuticorin. From cavities in coral.
Arca fusca, *Brug.* Rámésvaram. From cavities in coral.
,, (Scapharca) sp. Tuticorin.
Mucula layardi, *Ad.* Rámésvaram and Tuticorin.
Leda mauritiana, *Sow.* Rámésvaram and Tuticorin.
Pecten crassicostatus, *Sow.* Rámésvaram and Tuticorin.
,, tendineus. Tuticorin.
,, varius, *Linn.* Rámésvaram.
,, (Pallium) plica, *Linn.* Rámésvaram.
Spondylus, sp. Rámésvaram.
Vulsella rugosa, *Lmk.* Rámésvaram. From cavity in coral.
Ostræa, sp. Rámésvaram.

PISCES.

The following list comprises those species of fishes which I have either recorded or preserved during my visits to Tuticorin or Pámban, which latter place I made my headquarters while exploring the coral reefs which fringe the shores of Rámésvaram and the neighbouring islands. These visits have always been made during the months of July and August, so that my examination of the fish fauna has been confined to a very limited period of the year, and it will doubtless be found, on more extended research, to vary according to the season or monsoon.

The most characteristic feature of the fauna, as contrasted with that of other parts of the coast of the Madras Presidency, is the prevalence of the so-called " coral fishes " [1] (*Chætodon, Heniochus, Pseudoscarus,* &c.), for the most brightly coloured fishes which abound over the reefs and

---

[1] Indicated by an asterisk.

feed either on the small delicate marine invertebrates which swarm on the living corals, or, if their teeth are adapted for the purpose, on the soft parts of mollusc, which they extract by gnawing or boring holes into the hard substance of the shell. As stated by Haeckel,[1] an explanation of the bright colouring of the fishes is found in the Darwinian principle, that the less the predominant colouring of any creature varies from that of its surroundings, the less likely it is to be seen by its foes, the more easily it can steal upon its prey, and the more it is fitted for the struggle for existence.

Conspicuous by their abundance were several species belonging to the family Sclerodermi, including *Balistes* (file or trigger fish), whose jaws are armed with sharp teeth, and which are said to be injurious to the pearl fishery by preying on the pearl oyster. Present, too, in great numbers, were several species of the family Gymnodontes, *Tetrodons* (globe or frog fishes), including the beautifully marked little *T. margaritatus* and *Diodons*, which have a bad reputation among the natives as being very poisonous.

ELASMOBRANCHII.

(Sharks and Rays.)

FAMILY CARCHARIIDÆ.

Carcharias.[2]—Several young species commonly met with in the fish markets.

Zygæna malleus.   Tuticorin and Pamban.   " Hammer-head."

FAMILY SCYLLIIDÆ.

Stegostoma tigrinum.   Tuticorin.
Chiloscyllium indicum.   Tuticorin.

FAMILY PRISTIDÆ.

(Saw-fishes.)

Pristis cuspidatus.   A specimen 18 feet in length brought on shore at Tuticorin in 1887.

FAMILY RHINOBATIDÆ.

Rhinobatus granulatus.   Tuticorin.

---

[1] *A Visit to Ceylon*. Eng. Transl., 1883.
[2] The synonymy of Day's *Fishes of India* and *Supplement to the Fishes of India*, 1888, is followed throughout.

FAMILY TORPEDINIDÆ.

Narcine timlei. Pamban. Electrical Ray.

FAMILY TRYGONIDÆ.

Trygon sephen. Tuticorin.
,, uarnak. Tuticorin and Pamban.

FAMILY MYLIOBATIDÆ.

Myliobatis nieuhofii. Pamban.

TELEOSTEI.

(Bony Fishes.)

FAMILY PERCIDÆ. (Perches.)

Lates Calcarifer. Tuticorin and Pamban. The " cock up " or " nair."
Serranus boenack. Tuticorin.
,, diacanthus. Pamban.
,, hexagonatus. Pamban.
,, fasciatus. Pamban.
,, salmoides. Tuticorin.
,, tumilabris. Tuticorin.
Lutianus decussatus. Pamban.
,, erythropterus. Pamban.
,, fulviflamma. Pamban.
,, rivulatus. Tuticorin and Pamban.
,, roseus. Pamban. " Red rock cod."
Therapon quadrilineatus. Pamban.
,, theraps. Pamban.
Pristipoma hasta. Pamban.
Diagramma crassispinum. Tuticorin and Pamban.
,, cuvieri. Pamban. See Day's Sup., p. 785.
,, griseum. Tuticorin and Pamban.
Scolopsis vosmeri. Tuticorin and Pamban.
* Apogon auritus. Tuticorin.
* ,, kalosoma. Pamban.
* ,, thurstoni. n. sp. Pamban. See Day's Sup., p. 784.
Cheilodipterus quinquelineatus. Pamban.
Gerres oyena. Pamban.

FAMILY SQUAMIPINNES.

* Chætodon auriga. Pamban.
* ,, collaris. Pamban.
* ,, vagabundus. Pamban.
* Heniochus macrolepidotus. Pamban.

* Drepane punctata. Very plentiful in Pamban fish market in July 1888.
Scatophagus argus. Tuticorin and Pamban.

FAMILY MULLIDÆ (Red Mullets.)

Upeneoides tragula. Tuticorin and Pamban.
Upeneus indicus. Pamban.

FAMILY SPARIDÆ (See Breams.)

Lethrinus karwa. Tuticorin.
,, nebulosus. Pamban.
Chrysophrys berda. Tuticorin. " Black rock cod."
Pimelepterus cinerascens. Tuticorin and Pamban.

FAMILY SCORPÆNIDÆ.

* Pterois miles. Pamban. " Flying dragon."

FAMILY TEUTHIDIDÆ.

Teuthis marmorata. Tuticorin.
,, oramin. Tuticorin and Pamban.

FAMILY BERYCIDÆ.

Holocentrum rubrum. Tuticorin and Pamban.

FAMILY KURTIDÆ.

Pempheris malabarica. Tuticorin.

FAMILY SCIÆNIDÆ.

Sciæna maculata. Tuticorin.

FAMILY ACRONURIDÆ (Surgeons.)

Acanthurus mata. Tuticorin and Pamban.
,, triostegus. Pamban.
,, velifer. Pamban.

FAMILY CARANGIDÆ (Horse-mackerels.)

Caranx gallus. Tuticorin and Pamban.
,, hippos. Tuticorin.
,, ire. Pamban.
,, rottleri. Tuticorin.
,, sansum. Tuticorin.
,, speciosus. Pamban.

Plate IV

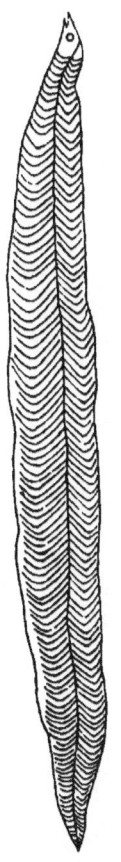

Leptacephalus

Platax teira. Pamban.
Lactarius delicatulus. Pamban.
Equula edentula. Pamban.

FAMILY SCOMBRIDÆ (Mackerels.)

Echeneis remora. Tuticorin. " The Remora."
,,   naucrates. Tuticorin.

FAMILY TRACHINIDÆ.

Sillago sihama. Tuticorin. " Whiting."

FAMILY GOBIIDÆ (Gobies.)

Gobius bynoensis. Tuticorin.
Gobiodon citrinus. Tuticorin.
Periophthalmus koelkreuteri. Pamban.
Boleophthalmus boddaerti. Pamban.

FAMILY BLENNIIDÆ.

Salarias marmoratus. Tuticorin.

FAMILY MUGILIDÆ (Grey Mullets.)

Mugil poicilus. Tuticorin.
,,   cunnesius. Tuticorin.
,,   speigleri. Pamban.

FAMILY CENTRISCIDÆ.

Amphisile scutata. Pamban.

FAMILY POMACENTRIDÆ.

* Glyphidodon antjerius. Tuticorin
*   ,,   cælestinus. Pamban.
*   ,,   notatus. Pamban.
*   ,,   sordidus. Pamban.
* Tetradrachmum aruanum. Tuticorin.
* Amphiprion sebæ. Pamban.

FAMILY LABRIDÆ (Wrasses.)

Cheilinus chlorurus. Pamban.
* Platyglossus dussumieri. Pamban.
* Pseudoscarus chrysopoma. Tuticorin and Pamban.
*   ,,   rivulatus. Pamban.

FAMILY PLEURONECTIDÆ. (Flat fishes.)

Plagusia marmorata. Pamban.
Cynoglossus macrolepidotus. Pamban.

FAMILY SILURIDÆ.

Arius thalassinus. Tuticorin and Pamban.

FAMILY SCOPELIDÆ.

Saurida tumbil. Pamban.

FAMILY SCOMBRESOCIDÆ.

Hemiramphus xanthoplerus. Pamban.

FAMILY CLUPEIDÆ. (Herrings.)

Pellona leschenhaulti. Pamban.

FAMILY MURÆNIDÆ. (Eels.)

Muræna tessellata. Tuticorin and Pamban.
,, undulata. Tuticorin.

FAMILY SYNGNATHIDÆ. (Pipe fishes.)

Syngnathus serratus. Tuticorin and Pamban.

FAMILY SCLERODERMI.

Balistes vetula. Tuticorin. "File fish."
,, mitis. Pamban. ,, ,,
Triacanthus strigilifer. Pamban.
Ostracion cornutus. Pamban. "Coffer fish."
,, nasus. Pamban. ,, ,,
,, turritus. Pamban ,, ,,

FAMILY GYMNODONTES.

Tetrodon hispidus. Pamban.
* ,, margaritatus. Pamban.
* ,, immaculatus. Pamban.
Diodon hystrix. Pamban.
* ,, maculatus. Pamban. See Day's Suppt., p. 809.

Leptocephalus, sp. (Pl. iv.)

As regards the curious pellucid *Leptocephali*, of which I have obtained a few specimens in the Gulf of Manaar, and a large number from the meshes of the fishermen's nets at Gopalpur, where they are known as sea-leeches, Dr. Günther says :[1]

"We must come to the conclusion that these leptocephatids are the offsprings of various kinds of marine fishes, representing, not a normal stage of development (larvæ), but an arrest of development at a very early period of their life; they continue to grow to a certain size without corresponding development of their internal organs, and perish without having obtained the characters of the perfect animal."

---

[1] *Introduction to Study of Fishes*, 1880, pp. 179-182.

VII.—INSPECTION OF CEYLON PEARL BANKS.

VII.—INSPECTION OF CEYLON PEARL BANKS.

HAVING received permission from His Excellency Sir Arthur Gordon, K.C.M.G., to accompany Captain Donnan, the Inspector of the Ceylon Pearl Banks, on his annual inspection cruise, I left Madras for Colombo by S.S. *Rewa* on the 3rd October 1889, taking with me some young plants of *Victoria regia*, reared in the nursery of the Madras Agri-Horticultural Society, for planting in the new Fort Gardens at Colombo. Some seeds of the *Victoria*, which had been sent from Madras earlier in the year, had germinated a short time before my arrival, and the young plants looked thoroughly healthy, so that it is to be hoped that the introduction of the water-lily will be successful.

While in Colombo I took the opportunity of examining the excellently preserved specimen of *Rhinodon typicus* in the Ceylon Government Museum for the sake of comparison with the specimen, more than 20 feet in length from the end of the snout to the extremity of the tail, which was cast on shore at Madras in February 1889, when I was unfortunately far away from head-quarters, so that the chance was missed of examining its stomach contents and internal anatomy. As the following extract shows, but few specimens of this monster Elasmobranch have been recorded [1]:—

"For many years the sole evidence of its existence rested upon a stray specimen, 15 feet in length, which was brought ashore in Table Bay during the month of April 1828, and fortunately fell into the hands of the late Sir Andrew Smith, then resident in Capetown, who named, described, and figured it. The specimen itself was preserved by a French taxidermist, who sold it to the Paris Museum, where it still remains in a much deteriorated condition. Forty years later, in 1868, Dr. Percival Wright, whilst staying at Mahé with Mr. Swinburne Ward, then Civil Commissioner of the Seychelles, met with this shark, and

---

[1] In his "*Account of the Pearl Fisheries of Ceylon*" Captain Steuart records having seen on one occasion "a spotted shark of a most fearful size; it was accompanied by several common sized sharks, and they appeared like pilot fish by its side."

obtained the first authentic information about it. It does not seem to be rare in this Archipelago, but is very seldom obtained on account of its large size and the difficulties attending its capture. Dr. Wright saw specimens which exceeded 50 feet in length, and one that was actually measured by Mr. Ward proved to be more than 45 feet long. Nothing more was heard of the creature until January 1878, in which year the capture of another specimen was reported from the Peruvian coast near Callao; finally, in the present century, Mr. Haly, the accomplished Director of the Colombo Museum, discovered it on the West Coast of Ceylon, and succeeded in obtaining two or three specimens. One of those was presented by that institution to the Trustees of the British Museum, and, having been mounted by Mr. Gerrard, it is now exhibited in the Fish Gallery, where it forms one of the most striking objects, although it must be considered a young example, measuring only 17 feet from the end of the snout to the extremity of the tail.

"A true shark in every respect, *Rhinodon* is distinguished from the other members of the tribe by the peculiar shape of the head, which is of large size and great breadth, the mouth being quite in front of the snout, and not at the lower side, as in other sharks. Each jaw is armed with a band of teeth arranged in regular transverse rows, and so minute that, in the present specimen, their number has been calculated to be about 6,000. The gill openings are very wide; and three raised folds of the skin run along each side of the body. Also in its variegated coloration this fish differs from the majority of sharks, being prettily ornamented all over with spots and stripes of a buff tint."

After waiting for several days on the chance of a moderation of the prevailing south-west wind, I left Colombo with Captain Donnan on the barque *Sultán Iskander*, which towed after her the diving boats, each with its crew composed of coxswain, rowers, divers, and mundueks who attend to the divers, letting them down by ropes, pulling them up, &c. The steam-tug *Active* followed us on the following day. As an inspection of the reported pearl bank off Negombo was out of the question owing to the heavy swell, we sailed straight on to Dutch Bay, where we anchored, after a somewhat boisterous journey, on the following morning, inside the long and rapidly extending spit of sand, which forms the western boundary of the bay, on which the sale bungalow, kottus, &c., were standing during my last visit in March at the time of the collapse of the pearl fishery. The Bay now presented a very deserted appearance, a few fishermen, living in huts and earning a modest living by curing sharks and bony fishes, and a number of natives, from near and distant parts of the island, engaged in searching for stray pearls in the sand formerly occupied by the washing kottus, the site of which was indicated by the remains of the fences and heaped up piles of oyster shells, and gaining as the reward of their labour from one to two rupees a day, being the sole human occupants of the sandy shore, on which hosts of wading birds were congregated. It was reported that one woman had found five pearls, each of the size of an ordinary pepper seed, for which she had been offered and refused 150 rupees.

The seaward face of the sand-spit was strewed with coral fragments rolled in by the waves from the reef, which intervenes between the shore and the pearl bank, and is partially laid bare at low tide; and the sand was riddled with the burrows of a very large *Ocypod* (*O. platytarsis*). If one of these crabs is killed and left on the shore, its fellow creatures carry it away into a burrow, and, doubtless, devour it.

On the day after our arrival at Dutch Bay we sailed in one of the diving boats to Karaitivu and Ipantivu islands and the mainland in search of a possible spot adapted for the requirements of a pearl camp at the next fishery. In the shallow water near the shore of Karaitivu island fishes—*Mugil* and *Hemiramphus*—some of which leaped into the boat and were eventually cooked, fell easy victims to fishing eagles and gulls. Two hauls of the dredge in the sand and mud brought up *Amphioxus*, *Lituaria phalloides*, the Trepang *Holothuria marmorata*, *Astropecten hemprichii*, *Philyra scabriuscula*, *Chloeia flava*, and many molluscs; the majority of the species of mollusc, both here and in Dutch Bay, being common to the Indian and Ceylon Coasts of the Gulf of Manaar. On the mainland forming the eastern boundary of Dutch Bay, into which the river Kala Oya discharges its water by several mouths, dense jungle and swampy ground teeming with the mollusc *Pyrazus palustris* reach right down to the water's edge; and, as we walked along the shore, we came across solid evidence of the recent presence of elephants. We were told by a native that bears and wild pigs are so thick in the jungle that one trips over them as one walks along!

In 1868 large numbers of young pearl-oysters are reported to have been spread over a considerable extent of the muddy bottom of Dutch Bay in from one to two fathoms of water, but the situation was, evidently, not favourable for their healthy growth.[1]

The weather being unfavourable for the work of inspecting, we had to remain unwillingly in Dutch Bay, the days being spent in cruising about, and dredging in the shallow water. But on the 29th, as the wind had changed and the sea abated, we made a start for the neighbouring pearl bank—Muttuwartu Par—to which we were towed by the *Active*. As soon as we had anchored on the south end of the bank a diver was sent down from the ship's side in $6\frac{3}{4}$ fathoms, and brought up his rope basket containing plenty of healthy, living oysters, which, he reported, came away easily from the "rock" to which they were attached by their byssi.[2] At the fishery in March the divers complained of the difficulty in detaching the oysters; and the ease with which they can be gathered is considered a sign whether they are ripe for fishing or not, the byssus being said to begin to break away from the substance to which it adheres tightly in the early existence of the oyster after the 5th year.

The excellent system which is employed in the inspection of the Ceylon banks, and by which a thorough knowledge of the condition of the banks is obtained, is as follows. The inspection barque is anchored in a position fixed on the chart by bearings from the shore. The steam tug, towing a boat with buoys bearing flags on board, first lays out buoys in the north, south, east, and west at distances of $\frac{1}{4}$, $\frac{1}{2}$, and $\frac{3}{4}$ of a mile from the barque. Buoys are then laid out at a distance of $\frac{3}{4}$ of a mile from the barque in the north-east, north-west, south-east, and south-west. Four diving boats, each with a coxswain in charge, 5 rowers, 3 divers, and 2 munducks, are arranged in line between the north $\frac{1}{4}$ mile buoy and the barque, the distance being equally divided between the boats. The rowers work round in a circle, and the divers make frequent dives in search of oysters until the starting point is reached. The boats are then again arranged in position, and the circle between the $\frac{1}{4}$ and $\frac{1}{2}$ mile buoys is explored. Lastly, the third circle, between the $\frac{1}{2}$ and $\frac{3}{4}$ mile buoys, is, in like manner, explored; so that, when

---

[1] E. W. H. Holdsworth. *Report on the Conditions and Prospects of the Pearl-Oyster Banks*, 1868.

[2] "The term rock is applied to pieces of coral, living or dead, averaging about a foot in diameter, which are scattered more or less thickly over certain parts of the banks." Holdsworth, *l.c.*

DIAGRAM. A

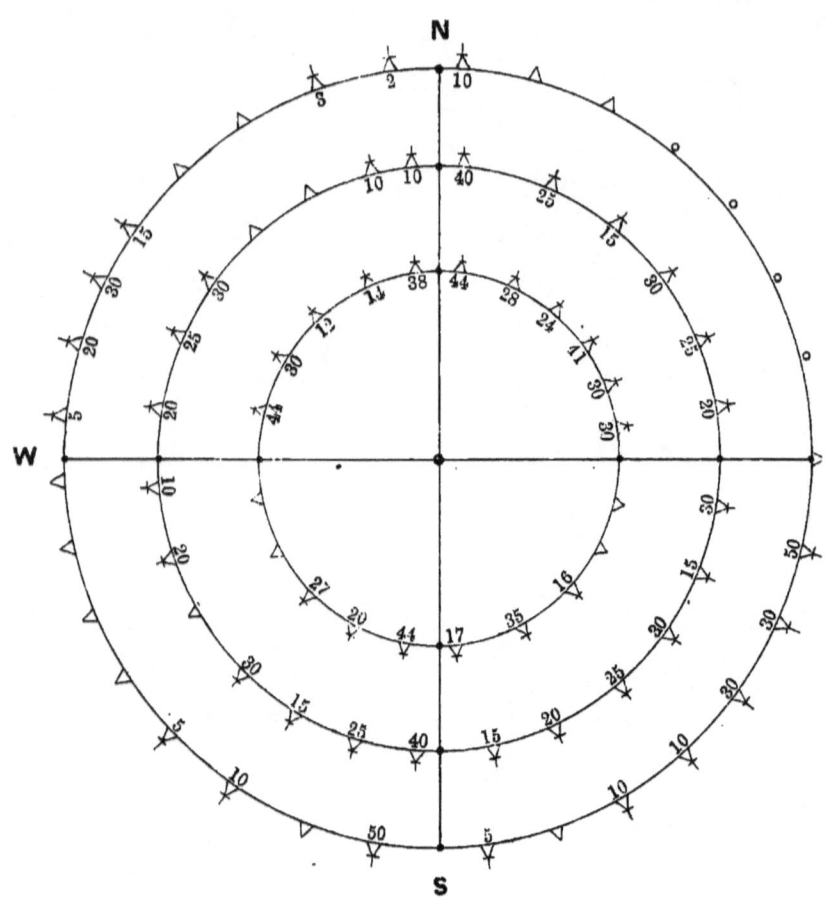

Lithographed in Central Survey Office, Madras
1890

DIAGRAM. B

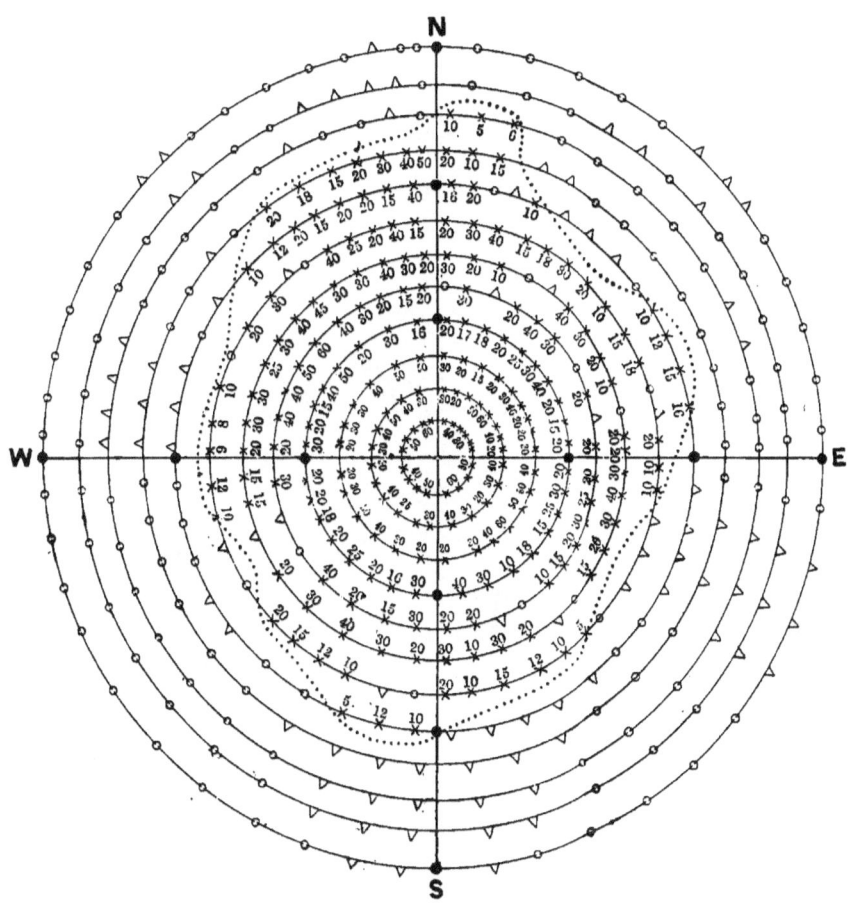

Lithographed in Central Survey Office, Madras
1899

this circle is completed, each boat has described three circles with the inspection barque as a centre. And, in this way, twelve circles in all are described by the four boats. The oysters are then brought to the ship, counted, and put in sacks daily, until a sufficient number (15,000) to form a sample for washing and valuation by experts has been collected.[1] The coxswain of each boat records on a diagram, provided by the Inspector, the approximate position of each dive which is made, the nature of the bottom (a triangle = rock, a circle = sand, and a cross = oysters), and the number of oysters lifted. Diagram A represents the day's work done by one boat over ground which, with the exception of a sandy patch between the north and east ¾ mile buoys, was rocky, and on which oysters were plentiful except over a portion of the outer circle. Diagram B, made up from the four coxswains' reports, represents a single day's work done by all the boats, and shows the distribution of the oysters over the area inspected, and the limits of the bank. As soon as the buoys have been taken up by the tug the inspection barque is moved to a new position 1½ mile distant from its former one, and the buoys are again laid out in circles, to act as guides to the boats in the next day's work. Without the assistance of the buoys the boats would not be able to describe separate circles, but would work in an irregular manner, and two or more boats would, very probably, go over the same ground. But, with the assistance of the buoys, the whole bank can be systematically surveyed.

The Muttuwartu Par, which was fished in the spring of 1889, is situated about 5 miles from the seaward shore of Dutch Bay, and covers an approximate area of 3 × 1½ miles, the depth of water over the bank ranging from 5 to 10 fathoms with an average of about 7 fathoms. The temperature of the water at the bottom, registered with a Negretti and Zambra's deep-sea thermometer, varied from 80° to 82° between 8 A.M. and 5 P.M. The specific gravity of the surface water, tested with a Twaddell's hydrometer regulated for a temperature of 84°, was, approximately, 1·025. Between the bank and the shore is a coral reef, the presence of which was indicated by the waves breaking over its outer face amid a prevailing calm, and by gulls resting on the coral blocks. The most conspicuous *Madreporaria*

---

[1] If a young bank is being inspected, samples are brought up by the divers, but they are not washed for valuation.

on this reef, which is surrounded by 4½ to 5 fathoms of water, belong to the genera *Madrepora* and *Pocillopora*, while *Galaxea* and *Leptoria* are present in less abundance. The bright white patches of sand, which cover large spaces between the coral growths, teem with *Protozoa* and a calcareous *Alga*, and are more rich in delicate molluscs than any other deposit which I have examined in the Gulf of Manaar. Sheltered among the coral tufts were sluggish *Holothurians* and hosts of small Crustaceans; and, clinging to the branches of a *Madrepore*, I found a single specimen of the quaint *Thenus orientalis*.

Outside the seaward face of the pearl banks on the Indian coast of the Gulf of Manaar the depth of the sea increases very gradually, so that, for example, outside the Tholayiram Par, a depth of only 15 to 20 fathoms is reached at a distance of 3 miles. Outside the Muttuwartu Par, however, the area of shallow water ceases very abruptly, and the depth increases rapidly to 150 fathoms at a distance of three-quarters of a mile from the seaward face of the bank, where the following temperatures were recorded:—

Surface	.. 83°	60 fathoms	.. 68°
10 fathoms	.. 81°	100 ,,	.. 61°
20 ,,	.. 80°	150 ,,	.. 55°
30 ,,	.. 72°		

Several hauls of the dredge brought up *Polytrema cylindricum*, *Gorgoniæ*, *Heteropsammia cochlea*, *Cirrhipathes spiralis*, *Spongodes* sp., *Fibularia ovulum*, &c., but no pearl-oysters.

The divers received instructions to keep apart for me everything, other than oysters, which they came across during their day's work, under the general heading of corals, shells, *poochers*, and weeds; and, by examination of the specimens which they reserved and going rapidly over the oysters, I was enabled not only to make a rich collection which awaits future investigation, but also to ascertain roughly in what respects the fauna of this portion of the West Coast of Ceylon differs from that of the Indian Coast of the Gulf of Manaar. The first day's inspection of the Muttuwartu Par showed not only that the oysters were very abundant, in spite of the disturbance to which they were subjected during the fishery in the spring, 4,580 living specimens being brought up in 291 dives; but, further, that the coral-incrusted shells, to which I have already referred

(p. 38), as being a distinguishing characteristic of this bank as compared with the Tholayiram Par, are very abundant, and belong to the genera *Madrepora*, *Montipora*, *Hydnophora*, *Porites*, *Pocillopora*, *Galaxea*, *Cyphastræa*, *Cæloria*, *Favia*, and *Goniastræa*; the living corals growing on the shells of living oysters, which, did they migrate, would have, sometimes, to carry about with them a weight of nearly 8 ounces. The coral-incrusted shells had, prior to the fishery of the Muttuwartu Par this year, only been seen by Captain Donnan on the North-west Chéval Par; and, when the oysters disappeared from the latter in 1888, the drift-oysters, which were eventually found, were recognised by the coral-growths upon them. Arborescent sea-weeds, forming tangled masses, such as abound on the Tholayiram Par, were conspicuously absent; but the oyster shells were largely encrusted with incrusting sponges, and the orange-coloured *Axinella donnani*, which receives its specific name after the present Inspector of Pearl Banks, was very common. In addition to the shell-incrusting corals massive corals, mainly belonging to the genus *Madrepora*, flourish on the bank, forming a convenient habitat and hiding place for Annelids, Crustaceans, Molluscs, &c., which can live there safe from the attacks of predaceous enemies. The sea bottom is, so far as I could gather from repeated examination, on different parts of the bank, of the residue left after shaking up the oysters in a bucket of water, and of the contents of the digestive tract of a Holothurian (*H. atra*) which abounds on the bank, mainly composed of a white deposit, such as I have only seen on the Indian Coast of the Gulf of Manaar, which is composed of a calcareous *Alga* and *Foraminifera*, among which *Rotalia calcar*, *Heterostegina depressa*, and *Amphistegina lessonii* are the most conspicuous. It was long ago pointed out by Captain Steuart that the places, on which pearl fisheries have been successfully held in Ceylon, appear to be beds of *Madrepore* of irregular heights, having the spaces between the ridges nearly filled up with sand. The transparent clearness of the water over the banks, and the clean state of the sea bottom, which is free from sediment carried down by currents, must, I think, be regarded as important conditions favouring the healthy growth of the oysters thereon.

Swimming about on the surface of the water over the bank were many black and yellow striped sea-snakes, which are believed by the divers to feed on the oysters. Indeed, in 1862, the European diver reported that he had seen the snakes eating the oysters, darting into the shells when opened. But this report must be viewed with grave suspicion. Apart from snakes, the reputed enemies of the pearl-oyster on the Ceylon banks are molluscs, fishes, and currents. Among molluscs are mentioned the Chank (*Turbinella rapa*) and a big *Murex* (*M. anguliferus*), known as the Elephant chank. But, as Mr. Holdsworth observes, "they may be looked on as part of the vermin of the banks, but I have no reason to think they cause more destruction on the oyster beds than the hawk and the polecat do among the game of an ordinary preserve." It is noticeable that the little *Modiola* known as *suran*, which assumes such a prominent position in the reports of the Inspector of Pearl Banks at Tuticorin, does not, though present, occur, so far as I am aware, in any great quantities on the Ceylon banks. Among fishes the trigger fishes (*Balistes*), commonly known as "Old Wives", are abundant on rocky parts of the banks, and I saw many specimens caught by the boatmen fishing from the side of the ship as we lay at anchor. Concerning these fishes Captain Steuart reports that "the sea over the pearl banks is well stocked with various fishes, some of which feed on the oysters, and, when caught by the seamen on board the guard vessel, pearls and crushed shells are often found in their stomachs, particularly in the fish called by the Malabars, the Clartee; by the Singhalese, the Pottooberre; and by seamen, the Old Women. This fish is of an oval-shape, about 12 inches in length and 6 inches in depth from the top of the back to the under part of the belly, and is covered with a thick skin. We saw ten pearls taken from the stomach of one of these fish on board the *Wellington*." The contents of the stomach and intestines of *Balistes*, which I examined while we were inspecting the Chéval Par, consisted entirely of young oysters crushed by their sharp cutting teeth. In addition to the trigger fishes, Rays are said to be always more or less numerous on the banks, and Mr. Holdsworth states that "when the fishery of 1863 commenced on the south-east part of the Chéval Par, the divers reported the ground so covered with skates as to interfere with their picking up the oysters. After a day or two the continual disturbance by the divers had the effect of driving the skates away from that part of the bank, and these fish, many of them of very large size, were seen going in the direction of the Modrigam, which was then covered with oysters, whose age was estimated by the Superintendent at $2\frac{1}{2}$—3 years, by the Inspector at $3\frac{1}{2}$—4, and by the native headman at 4 years. The skates were in shoals, and their total number was estimated at from 10 to 15 thousand. Further, in his report on the inspection of banks in March 1885, Captain Donnan notes the fact that "on the way from the north Mótaragam, and just about the south side of the bed of oysters, we passed through a large patch of thick discoloured water, caused by a shoal of Rays plundering about on the bottom, and stirring up the sand. Some of them could, at times, be seen near the surface, and I have no doubt they were feeding on the oysters." Some years ago the Sea Customs Officer at Dutch Bay counted as many as 300 Rays in a single haul of a fishing net. The native belief is that the Rays break up the oyster shell with their teeth, and suck out the soft animal matter. The stomach of a big Ray (*Ætobatis narinari*), 5 feet in breadth and with a tail $8\frac{1}{2}$ feet in length, which was caught by fishermen from a canoe off Silávaturai when we were at anchor there, consisted of sea weed. The same fishermen caught for me off the Silávaturai reef a male Dugong, 9 feet in length, whose stomach contents consisted of sea weed and large numbers of a *Nematoid* worm.

It was roughly estimated as the result of the inspection of the Muttuwartu Par, which lasted over three days, an average of 589 yards and 16 oysters to a dive being allowed, that it contained 30 million oysters spread over an area of $9\frac{1}{2}$ million square yards, which should produce a revenue of 5 lakhs of rupees.

On November 2nd we left the Muttuwartu Par, and anchored in 8 fathoms, about 2 miles further north, so as to hunt for a possible bed of oysters. The divers, making the usual preliminary dives, brought up blocks of dead coral-rock with living *Turbinariæ* and *Porites* growing on them, and containing, imbedded in the crevices, a large number of *Foraminifera*. The sample of 15,000 oysters from the Muttuwartu Par, which were beginning to be unpleasant fellow-passengers, was sent up to Silávaturai to be washed. It is stated by Cap'.ain Steuart that the offensive effluvium of decomposing oysters "is not considered to have an unhealthy tendency on the persons engaged in the kottus, and it is astonishing how soon the most sensitive nose becomes accustomed to the smell. Indeed some Europeans have fancied their appetites sharpened by visiting the kottus, and being surrounded by immense heaps consisting of millions of oysters in all stages of decomposition."

The surface of the water, always rich in organisms, was exceptionally so on the following morning, the tow-net, dropped from the stern of the barque and kept distended by the gentle current which was running, becoming speedily filled with a gelatinous mass composed mainly of *Sagittæ* mingled with a host of *Ctenophora*, glassy molluscs, and hungry fishes preying on Crustacean and other larvæ. Only a few young oysters being found, we again proceeded northward, and anchored in 8½ fathoms, the preliminary dives bringing up *Madrepores* with *Antedons* entwined round their branches, and large *Melobesian* nodules. Again only a few scattered oysters were obtained as the result of a day's work, but the divers brought me many specimens of *Alcyonians*, and the bright-red sponge *Axinella tubulata*, living attached by a broad base to dead coral-rock, and associated with its commensal worm.[1] The following temperature observations were made half a mile west of the ship, where no bottom was reached with the sounding line at 140 fathoms:—

Surface	.. 81·5°	50 fathoms	..	75°
20 fathoms	.. 76·5°	100 ,,	..	62·5°
30 ,,	.. 76°	140 ,,	..	55°

On the afternoon of the 4th, we moved on, still northward, to the Karaitívu Par,[2] which was estimated, at the inspection in November 1887, to contain 1,605,465 oysters. The divers, going down from the ship, alighted on a bank of *Fungiæ*, and brought up some living 5-year old oysters and *Melobesium* nodules. Attached to one of the nodules was an extensive creeping colony of the delicate crimson-coloured organism named *Tubipora reptans* from the single small specimen which has hitherto been recorded by Mr. H. J. Carter.[3] The present specimens were in a more advanced stage of growth than the one described by Mr. Carter, which I examined in the Liverpool Museum last year, and the calycles were proportionately higher. By about four hours' work next morning a sample of 8,000 oysters was collected

---

[1] Vide *Ann. Mag. Nat. Hist.*, Feb. 1889, p. 89.
[2] The Karaitívu Par was fished in December 1889; but the fishery came to an abrupt termination owing to a diver being killed by a shark. Apparently three men went down into the water, and two came up almost directly, saying that the third had been carried off by a shark. The rest of the divers could not be prevailed on to resume work, and left the bank.
[3] *Ann. Mag. Nat. Hist.*, June 1880, p. 442.

for valuation, and the abundance of oysters may be judged from the fact that, on more than one occasion, as many as 100 oysters were brought up at a single dive. My own share of the morning's work consisted of a *Fungia* (*F. repanda*) and three living specimens of the pearl-oyster *Avicula* (*Meleagrina*) *margaritifera*, attached by its byssus to coral-rock. Captain Donnan informs me that he has only seen about a dozen specimens of this mollusc during his 28 years' experience as Inspector of the banks, so that it cannot be present in any abundance. Shell-incrusting corals, though present on the bank, were far less common than on the Muttuwartu Par.

On the afternoon of the 5th we sailed about 20 miles north, and anchored in 2 fathoms, 3 miles south of the village off Aripu, off Silávaturai, which is made the headquarters at times when the Chéval and Mótaragam (Mudrigam) banks are fished. Rising from the sandy shore between Aripu and Silávaturai is a miniature sand-cliff, reaching a maximum height of about 12 feet, and extending over a distance of about half a mile, which contains a thick bed composed almost entirely of pearl-oyster shells—evidence of the enormous number of oysters which have been taken from the neighbouring banks at fisheries in the past. Similar beds of oyster shells were exposed in sections nearly a mile inland. The Chéval and Mótaragam banks are situated from 9 to 12 miles out at sea in water varying in depth from 6 to 10 fathoms. Between the shore and the banks the water gradually reaches a depth of 6 fathoms; but, as in the case of the Muttuwartu and Karaitivu Pars, the depth increases rapidly to 150 fathoms outside the banks. The sea bottom between the shore and the banks is made up mainly of sand with many worn shells, a luxuriant growth of sea-weeds, and scattered coral patches. Among mollusca *Modiola tulipa* in an advanced stage of growth, and the chank (*Turbinella rapa*) were very abundant. No fishing for chanks is permitted south of the Island of Manaar, lest, at the same time, raids should be made on the pearl banks. The fishery is, however, actively carried on north of the island on a different system to that which is in force at Tuticorin (p 33), the boat-owners paying a small sum of money annually to Government, and making what profit they can from the sale of the shells.

Writing of the banks off Aripu, which have been, for many years, the sheet-anchor of the Ceylon fishery, Captain

Steuart observes that " the number of successful fisheries obtained on the banks lying off the Aripu coast, more than on any other banks in the Gulf of Manaar, and the high estimation in which the pearls from these fisheries are deservedly held, would seem to indicate some peculiar quality in the bottom of the sea in these parts, which is favourable to the existence of pearl-oysters, and for bringing them to the greatest perfection. We know there is something in the nature of the bottom of certain parts of the sea, which is favourable to the subsistence and growth of particular fishes, and which improves the flavour for the food of mankind: for instance, the sole and the plaice caught in Hythe bay on the Kentish coast are esteemed better than those caught off Rye on the western side of Dungeness; and we also know that cod, turbot, oysters, and, indeed, most edible fishes are prized in proportion to the estimation in which the banks are held, from whence they have been taken."

In 1885 Captain Donnan attempted to cultivate the pearl-oyster on a coral reef, three miles from the shore, which was considered to be sufficiently far removed from the baneful influence of the Aripu river during the freshes. A tank for the reception of the oysters was dug in the centre of the reef, and surrounded by blocks of coral to form a barrier round its edge, heaped up high enough to be just awash at the highest tide. But the experiment failed, as, out of 12,000 oysters which were placed in the tank, only 27 remained alive at the end of seven months. " Some of the oysters," Captain Donnan writes, " may have been washed out of the tank by the south-west monsoon sea, as it was not completely sheltered from the wash of the waves, but the bulk of them have, I believe, died off and been destroyed by some fish preying upon them. About 100 dead shells were found in the bottom of the tank, many of which bore evidence of having been bored and nibbled away. It is just possible that some fish may have got into the tank, and preyed upon the oysters, either by getting over the coral barrier around it, which would be slightly under water at high-water, or through the interstices of the coral underneath. The experiment so far has been a failure, and may be attributable to four causes :—

" (1) overcrowding the oysters in the tank ;
" (2) deficiency of nourishment in water so near the surface ;

" (3) destruction by fish, which had got into the tank, and preyed upon them;

" (4) by excessive agitation of the water in the tank during the south-west monsoon sea; or, probably, to all these causes combined."

In March 1886 another experimental tank was made on a more sheltered part of the reef, and 5,000 oysters were placed in it. But, in the following year, all the oysters were found to be dead. The artificial cultivation of the pearl-oyster was attempted some years ago in a nursery made in the shallow muddy water of the Tuticorin harbour without success; and, in his final report to the Ceylon Government, Mr. Holdsworth expresses his opinion that there is no ground for thinking that artificial cultivation of the pearl-oyster can be profitably carried out on the Ceylon coast, as the conditions necessary for the healthy growth of the oysters are not to be found in the very few places, where they could be at all protected or watched.

On the way to Captain Donnan's tank, which we visited, we rowed over extensive banks of *Alcyonians*, of the luxuriant growth and size of which only a very feeble idea is obtained from specimens as seen in museums. On the sandy bottom a large number of Echinoderms, solitary or clustered together, were clearly visible; and, with the assistance of the divers and the dredge, the following species were procured:—*Temnopleurus toreumaticus*, a violet-spined *Temnopleuroid*, *Oreaster thurstoni*, *Salmacis bicolor*, *Laganum depressum*, *Fibularia volva*, *Echinolampas oriformis*, *Holothuria atra*, and *Colochirus quadrangularis*. These species, as also *Oreaster lincki* and *Linckia lævigata*, which abound on the Muttuwartu Par, are all found on the opposite coast of the Gulf of Mannar. A single young specimen of *Hippocampus* was also brought up in the dredge. The tank, washed by the gentle swell, showed no signs of pearl-oysters, which had, doubtless, been smothered and disappeared below the surface of the bottom. But growing from the inner side of the barrier of dead coral which formed the wall of the tank was a fringe of living corals—*Montipora*, *Pocillopora*, *Madrepora*, &c. As these corals had grown in their present position since the construction of the tank, which was built up entirely of *dead* blocks of solid coral brought from the shore, the living corals on the reef being found to be too brittle to form a suitable wall, it was obvious that, as the tank was built in March 1886, the age of the corals did not exceed three years and nine months. Accordingly I had the largest specimen of *Montipora* carefully detached from the dead coral-rock on which it was growing, and found that it measured 40 inches in length, 9 inches in height, and 16 inches in breadth, and weighed 17 pounds.

After remaining at anchor for some days off Silávaturai, we started on the morning of the 10th for the western side of the great Chéval Par, which is known by the divers as *kodai* (umbrella) Par from the prevalence on it of a shallow cup-shaped sponge, *Spongionella holdsworthi*, which is supposed, by their imaginative brains, to resemble an umbrella. In a letter to Mr. Bowerbank, by whom this sponge was described,[1] Mr. Holdsworth stated that " is only found on the 9-fathom line of the large pearl bank. It is attached to pieces of dead coral or stones. When alive it is of a dark brown; and when taken out of water it looks exactly like dirty wet leather.... This sponge is so strictly confined to the locality above mentioned that its discovery by the divers is considered the strongest evidence that the outer part of the bank has been reached." Another conspicuous sponge on this bank was the large, pale pink-coloured *Petrosia testudinaria*, which also lives on the Tholayiram Par.

It was from the Chéval Par that, in 1888, about 150 millions of oysters, ripe for fishing, disappeared in the space of two months, between November and February. This disappearance *en masse* was attributed by the natives to a vast shoal of rays, called *Sankoody tyrica* or *Koopu tyrica*, which is said to eat up oyster shells. But the more practical mind of the Inspector of the pearl banks attributed the disaster—for such it was from a financial point of view—to the influence of a strong southerly current, which was running for some days in December—a current so strong that the Engineer of the *Active* had to let go a second anchor to prevent the ship from dragging.

The divers, going down from the ship as soon as we were at anchor over the bank in 6½ fathoms, reported abundance of young oysters, whose average breadth at the hinge was ·75 inch, said by some to be three months, by others six months' old, and brought up samples, from the rocky bottom interspersed with patches of fine sand, attached to dead coral,

---

[1] *Proc. Zool. Soc.*, 1873, p. 25, pl. v.

Melobesiæ, sponges, and any other rough surface suitable for the attachment of the byssus. That the pearl-oyster prefers a rough to a smooth surface as an anchorage is shown not only by its usual habitat, but also by the observation that young oysters have been found clinging to the coir rope moorings of a bamboo, but not to the bamboo itself or the chain moorings. The number of young oysters on a small nodule brought up by the divers was counted, and found to be 180, scattered among which were 20 specimens of the little *Suran*.

The prevailing stony corals on the west Chéval Par, brought up by the divers with dense clusters of young oysters adhering to them, belonged to the genera *Porites*, *Astræa*, and *Cyphastræa*, growing from a base of conglomerated sand-rock, which is known by the divers as "flat rock." These corals, when broken up, proved a rich hunting ground for small crustaceans, tubicolous worms, and lithodomous mollusca. Very abundant on the bank were the bright-red *Juncella juncea* and the corklike *Suberogorgia suberosa*, on the axes and branches of which clusters of oysters were collected.

At the time of his last inspection of the west Chéval Par in 1888, Captain Donnan found a large portion of it stocked with oysters one year old, which had, in the interval between the inspections, died from natural causes or been killed off, and replaced by another brood. The life of the pearl-oyster must be a struggle ,not only during the time at which it leads a wandering existence on the surface,[1] and is at the mercy of *pelagic* organisms, but even after it has settled down on the bottom, where it is liable to be eaten up by fishes, Holothurians, molluscs, &c., or washed away from its moorings by currents; and comparatively few out of a large fall of "spat" on a bank can reach maturity even under the most favourable conditions. "Much," Captain Steuart writes, "appears to depend on the depth of water over the ground, and the nature and quality of the soil upon which broodoysters settle, whether any portion of them eventually reaches the age of maturity. If the deposit be of small extent, or be thinly scattered, the young oysters are often devoured by fishes, before the shells are hard enough to protect them. But, when the deposits settle in dense heaps upon places

---

[1] Young pearl-oysters have been found attached to floating timber and buoys, and to the bottoms of boats.

favourable for their nourishment and growth, many of them survive to become the source of considerable revenue." How great is the struggle of the pearl-oyster for existence is very clearly shown by the records of the Tuticorin inspections, in which, time after time, a bank is noted in one year as being thickly covered with young oysters, and in the next year as being blank. Not, in fact, till a bank is thickly covered with oysters two years' old can any hope be held out that it will eventually yield a fishery.

Outside the west Chéval Par a sand flat extends for some distance north and south, from which the dredge brought up masses of coarse, broken shells, and, among other specimens, large numbers of *Amphioxus* and *Clypeaster humilis*, and single specimens of *Ophiothrix aspidota* and *Astropecten hemprichii*; the digestive cavity of the latter being distended by a large *Meretrix* (*M. castanea*) and seven other smaller molluscs, which it had swallowed. From the stretch of sand between the east and west Chéval Pars the Echinoids *Echinodiscus auritus* and *Metalia sternalis* were obtained.

During our stay on the west Chéval Par, large numbers of the butterfly *Papilio* (*Menelaides*) *hector* were seen daily fluttering around the ship 10 miles out at sea. The *Active* steaming at the rate of 4 knots an hour, and the diving boats under sail caught many seir fish (*Cybium guttatum*) with a long line towing astern and made fast to the yardarm of the lug sail, and baited with a piece of white rag. For catching seir the hooks are, sometimes, baited with a small fish or the white of a coconanut cut into the shape of a fish. From the barque at anchor many *Balistes* and the crimson-coloured *Lutianus erythropterus* were caught by the crew with lines baited with fish. The stomachs of the former always contained crushed pearl-oysters, and those of the latter small fishes.

On the 14th we inspected the small Periya Par, situated 3 miles westward of the west Chéval Par, which we found irregularly stocked with young oysters. Sounding seaward from the bank, we found 9 fathoms at a distance of 1 mile, 14 fathoms at a distance of 2 miles, and did not strike bottom at 150 fathoms at a distance of 4 miles. The sea bottom shelves here less abruptly than outside the Muttuwartu Par, where a depth of 150 fathoms was obtained at a distance of $\frac{3}{4}$ of a mile from the seaward face of the bank. The thermometer registered 54° at 150 fathoms, and 59° at 100 fathoms, the surface temperature being 83°. On this and the two preceding days a bright blue-eyed *Palæmonid* larva was very abundant on the surface.

The next four days, during which the weather was very unpleasant and suggestive of a cyclonic storm in the Bay of Bengal, were spent in inspecting the east Chéval Par. The divers, going down as soon as we had anchored at the north end of the bank, brought up blocks of incrusted sand-rock, and specimens of the black-coloured sponge *Spongionella nigra*, but no oysters, which were, in fact, absent over the entire bank. This bank is mainly characterised by the abundant growth on it of *Suberogorgia suberosa*, on the branches of one of which an *Astrophyton* (*A. clavatum ?*) was entwined, and a heather-like *Hydroid*, the tangled branches of which were studded with the striped *Avicula zebra*, and which should afford good anchorage for young oysters. Conspicuous among other specimens, which were obtained, were the sponge *Hircinia clathrata* affording a home to *Balanus* (*Acasta*) *spongites*, the corals *Turbinaria crater* and *Turbinaria patula*, and the Echinoderms *Antedon palmata*, *Salmacis bicolor*, *Clypeaster humilis*, and *Echinaster purpureus*. A single specimen of *Ophiothrix aspidota* was found coiled up in a cavity in a block of *Porites*. As on the other banks which we inspected, sea-weeds were not present in any quantity. The quantity of weed on the banks is said, however, to vary much from year to year.

The inspection of the east Chéval Par completed, we went a short distance south, and spent a couple of days on the Mótaragam Pars, which were also blank so far as oysters were concerned. The pearls from these Pars are highly valued by the pearl-merchants, and, at the fishery of 1888, the oysters fetched from 100 to 109 rupees per thousand at auction, a single day's fishing realizing over 60,000 rupees. The weather had cleared up by this time, and the divers were again able to work in comfort for a short time. Rain interferes very much with an inspection, as the divers complain that it makes them cold and shivery when they come out of the water. Here, as on the east Chéval Par, the animal collected in greatest abundance was *Clypeaster humilis;* but the divers also brought up many specimens of the chank, the unpleasant looking animal of which is eaten by the natives; *Pinna bicolor*, which is said to occur on the sandy parts of the banks in beds of some extent; and the hammer-headed oyster. The *Hydroid*, which was so conspicuous a feature of the east Chéval, was absent from the Mótaragam Par.

At this stage a strong south-west wind came on, accompanied by an unpleasant swell, and drove us into Silávaturai; but, luckily, all the important work of the inspection tour was finished, two small banks alone remaining to be examined. A rolling journey on the tug *Active* brought me back to Colombo, and my second visit to Ceylon, more auspicious than the first, was over.

During the last quarter of a century, the Ceylon Government has derived a handsome profit from its pearl banks, which have been lucratively fished on nine occasions; while, during the same period, the banks belonging to the Madras Government have yielded only a single fishery, not because the oysters have ceased to settle, when young, on the banks, but because they have failed, owing to a combination of physical and other unfavourable conditions, to reach maturity there. Writing, in 1697, for the instruction of the Political Council of Jaffnapatnam, the then Commandant of that town justly remarked that the pearl fishery is an extraordinary source of revenue, on which no reliance can be placed, as it depends on various contingencies, which may ruin the banks, or spoil the oysters And this remark holds good after the lapse of two centuries. In 1740 the Baron von Imhoff, on his departure from the Government of Ceylon, in a memoir left for the instruction of his successor, stated that "it is now several years since the pearl banks have fallen into a very bad state both at Manaar and Tuticorin; this is mere chance, and experience has shown that, on former occasions, the banks have been unproductive even for a longer period than has yet occurred at present." And a century later, in 1843, Captain Steuart, at the commencement of his admirable "Account of the Pearl Fisheries of Ceylon," refers to the failure at that time of the now lucrative Ceylon fishery. Is it then rash, looking back to the fluctuating experience of the past, to express a belief that, in the not far distant future, the reputation of the Tuticorin banks will rival that of the at present well favoured banks of Ceylon?

The name of Captain Donnan has repeatedly appeared in this chapter, and I should be, indeed, ungrateful were I not to acknowledge not only the great assistance which I received from him in carrying out my zoological work, but also the vast store of information which I gathered from him during our month of banishment from the outside world.

www.ingramcontent.com/pod-product-compliance
Lightning Source LLC
Chambersburg PA
CBHW021917180426
43199CB00032B/357